What's Blooming A Guide to 100+ Wild Plants of Northwest Territories

Botanically accurate, but easy to use
Arranged by flower colour
Plants are trailside or roadside

Alexandra Milburn

Disclaimer

The author and publisher do not recommend that you use any of these plants for food or medicine. Text included on these topics is for information and entertainment only. If you do use these plants in any of the ways stated in this book, we are not liable in any way for any consequences.

Published by What's Blooming
whatsblooming@nt.sympatico.ca

Illustrations by Terrance Pamplin

Book design by Lapel Marketing & Associates Inc.

National Library of Canada Cataloguing in Publication Data

Milburn, Alexandra.

What's Blooming

Includes bibliographical references and index.

ISBN 0-9730819-0-2

1. Botany--Northwest Territories. I. Milburn, David, 1950- II. Title

QK203.N57M54 2002 581.9719'3 C2002-910515-3

Printed and bound in Canada
by Houghton Boston, Saskatoon

Acknowledgements

I am indebted to the instructors from the Athabasca University Wildflowers Course: Richard Dickinson for cheering me on when I first started this project, and to Dr. Trilochan S. Bakshi for critically reviewing the manuscript.

Appreciation also goes to the guys at Renewable Resources, Wildlife and Economic Development, especially Bob Bailey, Ken Caine and Gerry LePrieur, for their contributions to the development and completion of this book.

The author wishes to thank Gartner Lee Limited and MacDonald Environmental Sciences Ltd. for their gracious contributions in support of photo costs for this book.

A special thanks goes to Philip Gregory of Lapel Marketing for his clarity of vision in design and his persistence in completing this book.

To my sidekicks and loyal supporters, my husband David Milburn and our daughter Georgie, thanks for always being there.

I will be the gladdest thing
under the sun!
I will touch a hundred flowers
and not pick one.

Edna St. Vincent Millay

Dedication

To the memory of my brother,
Terry Porter, who loved the wilds
of Albert County, New Brunswick,
and now lies there in peace beneath the trees.

Foreword

Northerners tend to be travellers. We hit the trail with friends and family on snow machine, boat, car or truck. Sometimes we take a less mechanical approach and walk along trails, travel routes or just wander around town. Northerners also tend to be observant.

We have the opportunity to observe our plant neighbours when we are out in our own backyards, or stopping along the way at a highway campground or our favourite island hideaway. Has a plant along the trail caught your eye and you wonder what it is? Sometimes the kids will ask, "Hey, what kind of plant is that?" It is great to be able to respond with practised authority, "Oh, that's a fireweed."

Now with this book you can answer these questions and give the kids a better answer. Alexandra Milburn and her husband, David, have combined their talents to produce this very handy guide to more than 100 of the most common plants we will run across in our travels through the northern landscape. An overview of the ecozones of Northwest Territories heads up this useful volume, helping put plants in context with the landscape on which we and they live.

Alexandra has prepared carefully written descriptions of individual plant species, organized by flower colour, accompanied by David's wonderful full colour pictures. This presentation makes it easy to readily identify plants we find under foot or over our heads. The plant notes are rounded out with information on how these plants were used by Aboriginal and other people. Plant lore gives the reader a good sense of how valuable many plant species are and which ones to avoid.

Technical terms are kept to a minimum; however, botanically accurate names and a glossary are provided for those interested. This unique book has lots of information about many local species in an easily read and very functional format. This book would be useful for the family who travels our landscape, as a teaching aid for children as well as adults and a stepping stone for those pursuing a more academic approach to plant identification. Anyone at all interested in the plants we see outside everyday will find What's Blooming a valued companion on every trip we take. My copy is in the truck for quick reference next time I'm on the roadside or heading down the trail!

Bob Bailey

Bob has spent almost three decades in the North, working in the field of renewable resources. He has spent many hours on the land in work and recreation.

Author's Foreword

As a curious person, I have always had questions about the world around me. As a young girl, I would collect wild plants, but did not know how to identify them. There were no bookstores in the small town where I grew up, and at that time, the study of nature was left to the scientists. In my mid-twenties I finally found a field guide. It was difficult for me to use, because all the plants really do look alike to a novice. I had to start from scratch to figure out the structure of plants and try to determine which plant was which.

In 1982, my husband and I moved to Yellowknife. Living in an area of such natural beauty, with the wilderness and its wild things just at my doorstep, I began studying plants in earnest. The more I have studied, the more joy I have found in nature and the more peace I have found within myself. One of the greatest thrills for me is identifying an unfamiliar species. With a sense of accomplishment and closer communion with the natural world, I am set for the rest of my day.

In 1999, I began to write a book that would be accessible in terms of finding plants as well as for identifying and understanding them. Every plant in this book can be found by walking or driving. Botanical terms have been kept to a minimum. With this book, I hope you will be able to experience the same joy as I when you identify and learn about the northern wild.

Contents

Map 1. Ecozones of Northwest Territories

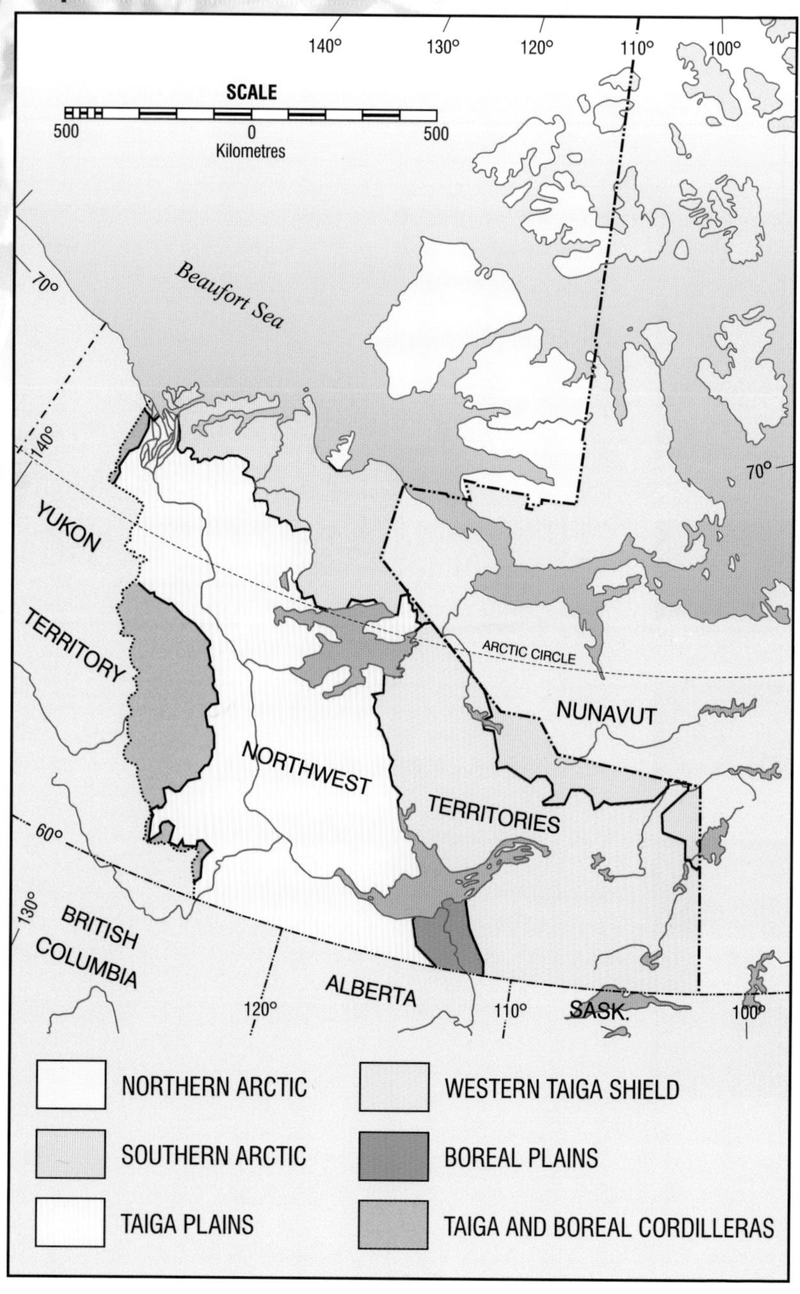

Map 2. Aboriginal Settlement Lands

Map 3. Species Collection Areas

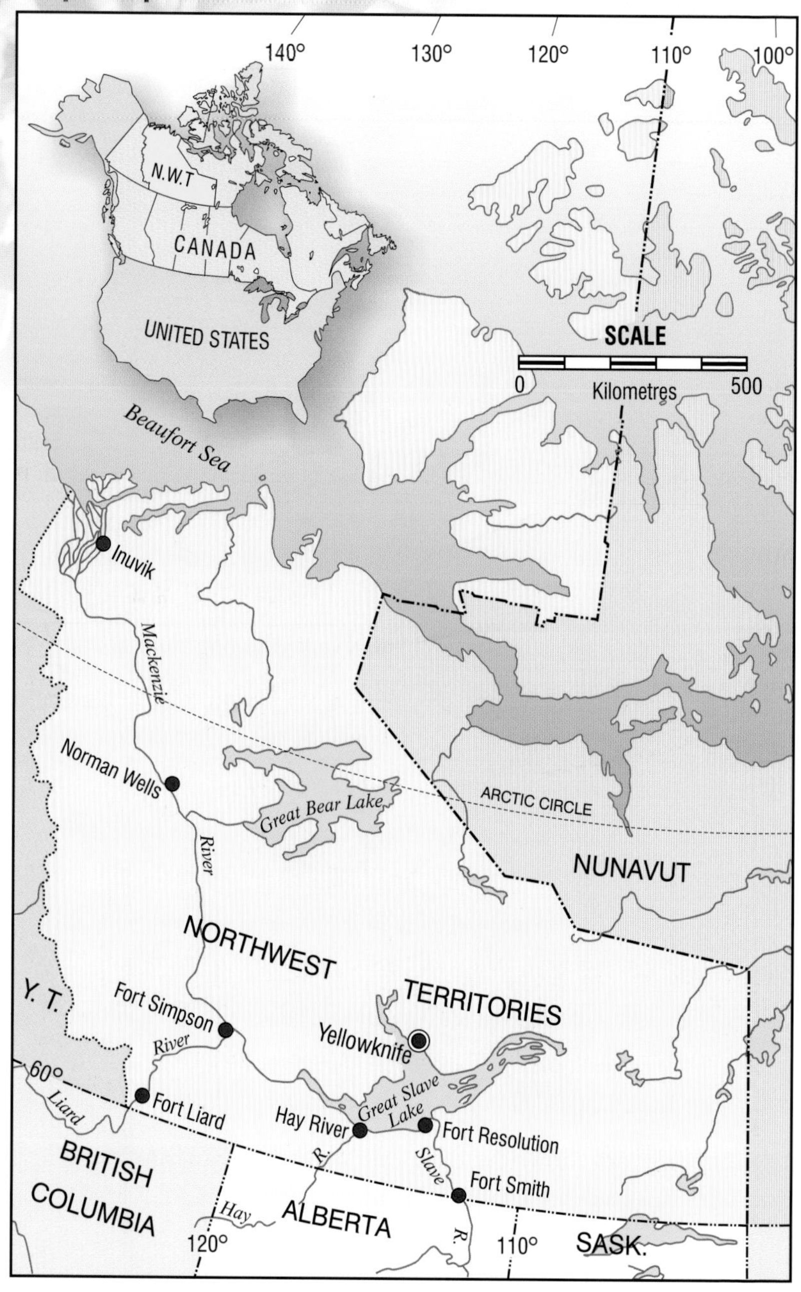

Introduction

Plants are everywhere in Northwest Territories. Even in areas, such as the tundra, that seem devoid of vegetation, there is a thriving diversity of plant life. We are blessed in this part of the world with a vast area of unspoilt natural beauty. The new territory of Northwest Territories came into being on April 1, 1999, and now comprises almost 1.5 million square kilometres.

Within this vast northern land exists the tenth largest river in the world, the fourth largest lake in North America and the tallest trembling aspen in the world. There also exists a diversity of cultures and traditions and a large number of political boundaries. This book concerns itself mostly with ecological distinctions (Map 1), although attention has been paid to the distinctions between traditional Aboriginal settlement areas (Map 2).

The plant collection areas are all on the beaten track, close to, or in, the main settled areas of NWT (Map 3). In an effort to make this book accessible to as many people as possible, only plants that can be found along the roadsides and walking trails have been included, with the exception of alpine azalea, which is a tundra species.

Sources

Wherever possible, information on Aboriginal plant use has been derived from original sources. In other cases, it was necessary to draw upon another author's interpretation. Please take the information provided as interesting and entertaining reading only. **We do not recommend that you use any of the plants in this book for food or medicine. Any such use is entirely at your own risk.**

The author has been able to use original research from sources, as listed in the bibliography, for the following Aboriginal groups: the Fisherman Lake Slave of the Deh Cho region, referred to in this book as the Dene of the South Slave or the South Slavey; the Dogrib people of Wha Ti; the Sahtu Dene; and the Inuit people of Baffin Island. Other sources used by the author are the interpretations of others, which the author has drawn upon, and may be from other areas of Canada or North America. Attribution is given in all cases where Aboriginal plant use is cited. For some plants, only western and historical references were available and in these cases, attribution is less specific, as it was in the sources from which information was derived.

Ecozones of Northwest Territories

The term ecozone is used to describe a landscape that has a number of common characteristics, such as climate, human activity, vegetation, soils, and geological and physiographic features. Within these larger ecozones are smaller ecoregions characterized by distinctive regional features. There are 15 ecozones in Canada. Six of these are within Northwest Territories and four are covered in this book: the Southern Arctic, the Taiga Shield, the Taiga Plains and the Boreal Plains (Map 1).

Southern Arctic

The Southern Arctic extends as a band along northern mainland Canada from the Richardson Mountains in Yukon to Ungava Bay in northern Quebec. This ecozone is commonly known as "the barrens" or "the barren lands," a name given to it by the first European explorers. The Southern Arctic, however, is not barren, but has the most extensive vegetative cover and highest diversity of species of the three arctic ecozones. Its climate of long, cold winters, short, cool summers, and minimal rainfall results in an extremely short growing season, particularly the farther north one goes. Because plants bloom, produce seeds and die within such a short time, the barrens seem to come alive all of a sudden with a diversity of wildflowers, shrubs, sedges and mosses. Typical shrubs include dwarf birch, willow, and heath species that are commonly associated with various herbs and lichens. Stunted spruce trees may be found along the sheltered banks of larger rivers. The landscape is an undulating terrain of hills and granite outcrops studded with many shallow lakes, ponds and extensive wetlands in lowland areas. Permafrost is continuous in this ecozone although warmer temperatures in summer will thaw the uppermost or active layer, providing much-needed moisture for plant life.

There are few human inhabitants in the Southern Arctic. The only communities of Northwest Territories in this ecozone are Tuktoyaktuk and Paulatuk. Nunavut communities include Kugluktuk, Chesterfield Inlet, Rankin Inlet and Arviat.

Taiga Shield

The Taiga Shield lies on either side of Hudson Bay and the western segment occupies portions of northern Manitoba, Saskatchewan, Alberta and Northwest Territories. This ecozone gets its name from the Russian word taiga or "the land of little sticks," that extends from the subarctic of Labrador to Alaska, and from Siberia to Scandinavia. Two very large yet distinctive areas comprise this ecozone: the Taiga Forest on the western fringe and the Canadian Shield in the central portion. Some of the world's oldest rocks are found in the Shield area just north of Great Slave Lake. The climate of this zone is considered subarctic, meaning that it has relatively short, but mild, summers and long cold winters. The broadly rolling terrain is dominated by large bedrock outcrops; many lakes, wetlands and open forests of stunted black spruce and jack pine dot the region. Low-lying areas are covered with peatlands and may remain saturated for long periods of time. The northerly limit of trees reaches well into the Taiga Shield. This ecozone has the largest concentration of long, sinuous eskers (glacial deposits of gravels and sands) in Canada. Permafrost is discontinuous but widespread. The combination of these unique features results in a diverse array of plant species including trees, shrubs, ferns and allies, graminoids and wildflowers.

Yellowknife, at the end of Highway 3, is the only major centre in the Northwest Territories portion of this ecozone. Along the highway south of Yellowknife, many of the species in this book can be found.

Taiga Plains

The Taiga Plains is an extension of the Interior Sedimentary Plain of North America. This ecozone is dominated by Canada's largest river, the Mackenzie, its many tributaries and two massive lakes: Great Slave and Great Bear. The area is bordered by the extensive Mackenzie Delta to the north, and the closed forests of the Boreal Plains in the south. Summers are short and cool; winters are long and cold. Because this ecozone is influenced so dramatically by cold Arctic air, snow and ice may last for up to eight months of the year. The landscape is characterized by broad lowlands and plateaus, widespread permafrost and poor drainage. Major rivers cut deep valleys in the limestone, shale and sandstone bedrock.

Trees such as black spruce are common within this ecozone. Along the nutrient-rich banks of larger rivers, white spruce and balsam poplar grow to sizes comparable to the largest in the boreal forests to the south. Many shrubs, such as dwarf birch, Labrador tea, bearberry, and willow grow here as well. Mosses and sedges are widespread and easy to find in this zone.

The major communities include Inuvik, Norman Wells, Hay River and Fort Simpson.

Boreal Plains

The Boreal Plains consist of a broad band extending from the Peace River country of northwest British Columbia to the southeastern corner of Manitoba, and extends into Northwest Territories only in the Slave River Lowlands. The word boreal comes from *Boreas*, the Greek god of the north wind. As with the other ecozones, summers are considered short and cool, and winters are long and cold. Because of its more southerly latitude, however, the extremes of climates found in the other ecozones are not experienced here. This ecozone differs from the others in Northwest Territories in that up to half of the area is covered by peatlands. The other dominant feature is the delta of the Slave River. Sporadic pockets of discontinuous permafrost with low ice content exist throughout this region. The vegetation is boreal, consisting of stands of trembling aspen, balsam poplar, jack pine, white and black spruce. Tamarack, black spruce, shrubs of the heath family and mosses cover the many poorly drained fens and bogs. Towering species of *Equisetum* can be seen along the Slave River Delta.

This region contains most of Wood Buffalo National Park, the largest park in Canada (44 840 km^2), and the Slave River Delta, one of the major freshwater deltas in Canada. It is also home to the whooping crane, perhaps Canada's most famous endangered species, and the world's largest free-ranging bison herd. The major communities include Fort Resolution and Fort Smith.

David Milburn

Characteristics of the Ecozones of Northwest Territories

Ecozone		Southern Arctic	Taiga Shield
Physiography		rolling tundra, rock outcrops and boulder fields, continuous permafrost	broadly rolling terrain, large granitic rock outcrops, many lakes and wetlands, discontinuous permafrost
Mean Temperature Ranges	Summer	4 to 6 °C	6 to 11 °C
	Winter	-18 to -28 °C	-11 to -25 °C
Annual Precipitation		200 to 500 mm	200 to 500 mm
Flora		dwarf birch, willow, heath plants, wildflowers, sedges, mosses, herbs and lichens	stunted trees, shrubs, ferns and allies, aquatic plants, grasses, sedges and wildflowers
Soils		numerous glacial deposits of sand and gravel; acidic; generally thin soils	acidic; generally thin soils

Southern Arctic

Taiga Shield

Taiga Plains	Boreal Plains	Ecozone	
broad lowland plains, deeply incised river channels, discontinuous permafrost	Large river delta, peatlands, pockets of discontinuous permafrost	Physiography	
7 to 14 °C	13 to 16 °C	Summer	Mean Temperature Ranges
-15 to -26 °C	-11 to -18 °C	Winter	
200 to 500 mm	300 to 400 mm	Annual Precipitation	
trees, shrubs, mosses, grasses sedges and wildflowers; most diverse flora in NWT	numerous coniferous and deciduous trees and shrubs, grasses, sedges and wildflowers, many unique species	Flora	
acidic; deposits or regions of thick soils especially nutrient-rich flats of large rivers	acidic; large deposits of thick soils along Slave River	Soils	

Taiga Plains

Boreal Plains

How to Use This Book

Plants are categorized by what the observer can readily see: seeds and flowers. Three categories have been used and entries are arranged alphabetically.

Category	Plants	Arrangement
Plants without seeds and without flowers	ferns and allies	alphabetical by common name
Plants with seeds but without flowers	gymnosperms or cone-bearing plants	alphabetical by common name
Plants with both seeds and flowers	grasses, sedges, rushes, aquatic plants, deciduous trees, shrubs and wildflowers	colour of flower and then alphabetical by common name

Fireweed, for example, is found in the third section of the book on p. 65, under Pink to Red Flowers, between dwarf raspberry and Indian paintbrush.

This book does not comprise all the plants in NWT, but a selection of 103. Because a plant is not included does not mean it has been overlooked; it simply means that I chose not to include it in this volume. Many of the plants are common; some are less common; and some are rare. I am confident, however, that you will see many of the species in your walks and drives. One of the true joys of the North is its summer, with the availability of wild, blooming plants everywhere you look.

It should be noted that the taxonomical reference for this book is A.E. Porsild and W.J. Cody's 1980 work, Vascular Plants of Continental Northwest Territories, Canada. Although some of the genus and species names may have gone out of fashion and been replaced, Porsild and Cody remains the standard reference for this part of the world. It is still widely available in bookstores and libraries. Family names, however, have been updated from that text.

WHITE FLOWERS

Wild Chamomile

Wild Chamomile — Sunflower Family

Matricaria maritima (ma-tri-KARE-ee-ah ... Y-mah) — *Asteraceae*

Matricaria means "dear mother," referring to the plant's great medicinal value. *Maritima* means "of the sea."

This cosmopolitan weed will be familiar to anyone who has ever traversed meadows or walked along the roadside. Growing to 50 cm or m... **d chamomile** has groove... s that may branch tow... e top of the plant. Each branch carries a flower at its tip. This plant is the seaside species that has adapted to drier habitats.

Leaves
Bright green; deeply and finely divided; appea... ge-like.

Flowers
One to several, ...isy-like; rays pure white; disks dark gold.

Plant Notes
Species of chamomile have been used since ancient times for their healing properties. Contemporary herbalists use it externally for wound healing and to treat inflammations. Internally, it is used for fever, indigestion, anxiety and insomnia.

In the well known children's story, Peter Rabbit's mother m... momile tea for him when he returned ... er MacGregor's garden. We can assume that he was suffering from not only indigestion, but also anxiety!

Chamomile was one of the plants used in Medieval times to make turf seats. During the days of the Tudors chamomile was planted for fast growing, aromatic lawns. A lawn of chamomile still grows on the grounds of Buckingham Palace.

Field Notes:

37

1 Plant category
2 Common plant name
3 Common family name
4 Botanical name of plant
5 Pronunciation of botanical name
6 Botanical family name
7 Derivation of botanical name
8 General description of plant
9 Distinguishing features
10 Food use, medicine lore or eccentricities of plant
11 Reader's notes

Wild Plants
without Seeds and without Flowers

Bristly Club Moss

Bristly Club Moss | **Club Moss Family**
Lycopodium annotinum (*Ly-ko-POAD-ee-um ann-oh-TEE-num*) | *Lycopodiaceae*

Lycopodium is Greek for "wolf foot," referring to leaf shape of other species; *annotinum* means "a year old," referring to the plant's growth.

Bristly club moss grows in open, dry, sunny places in low-arctic regions throughout the circumpolar world. A low-lying plant, bristly club moss looks like a small evergreen branch curving up from the forest floor. Its stems creep along the ground, forming mats up to 1 m in length. From the stems, small bundles of bristly tipped branches arise. Each branch terminates in a cone-like strobilus that contains the spores by which the plant reproduces.

Plant Notes

The club mosses are not really mosses, but vascular plants. True mosses do not have vascular tissue or strobili. Although small and of relatively insignificant size, club mosses grew as large as trees during the Carboniferous Period, 280 million years ago. Like horsetails (*Equisetum* spp., p. 6), club mosses contributed to the fossil fuel beds we use today. Club-moss was used historically as a fire-starter because the spores are rich in oil and are extremely flammable. At one time, the plants were used as flash powder for photography and theatrical effects. Club moss spores were used by the Cree to see the effect of a given illness. Spores were sprinkled on water; if they radiated toward the sun, the illness was not expected to kill the patient.

Leaves

Scale-like; pointing away from stem.

Field Notes:

Fragrant Shield Fern

Fragrant Shield Fern — **Fern Family**

Dryopteris fragrans (dry-OP-ter-is FRAY-grens) — *Polypodiaceae*

Dryopteris is from the Greek words for "oak," and "fern;" *fragrans* means "fragrant."

Fragrant shield fern can be found on cliffs and rock screes throughout our area. It may be confused with rusty woodsia (*Woodsia ilvensis*, p. 5) because both are densely tufted, but shield fern has curled dead fronds at its base. A stout rootstock holds the plant in the ground. Fronds have a dry, fragrant, woodsy smell.

Fronds

15-20 cm long; lance shaped; somewhat leathery; evergreen; chaffy on underside; pinnae with toothed edges.

Plant Notes

People of Baffin Island, Nunavut, traditionally used dried fronds to sweeten the smell of babies by putting bundles of dried ferns in the armpits. Older dried fronds were used in bundles as air fresheners. Tobacco mix was made from fronds as well.

An infusion made from the fronds of shield fern has been used since ancient times to keep the scalp healthy and make the hair shiny.

Field Notes:

Mountain-parsley

Mountain-parsley **Fern Family**
Cryptogramma crispa *(Krip-toe-GRAM-ma KRIS-pah)* *Polypodiaceae*
Cryptogramma is from the Greek *kryptos*, which means "to hide," and *gramma*, which means "a line." *Crispa* means "curled," referring to the leaf margins.

Fresh green in colour, **mountain-parsley** is an airy, leafy plant that grows in tufts in shallow organic soil and rock crevices north to Great Bear Lake. About 20 cm tall, mountain-parsley has both sterile and fertile fronds, which make it appear to be two separate plants.

Fronds

Sterile fronds look like parsley; fertile fronds are taller and erect, with rounded, spike-shaped leaves. Rolled leaf margins of fertile fronds hide line of spore clusters on underside of leaf segments.

Plant Notes

Ferns evoke an airy, romantic and verdant atmosphere. Although the habitat in the North is harsh and seemingly inhospitable, ferns still thrive and are welcome greenery amongst the grey, hard rocks of the Precambrian Shield. Mountain parsley can be elusive to those who overlook it as just two plants jumbled together.

Field Notes:

Rusty Woodsia

Rusty Woodsia **Fern Family**
Woodsia ilvensis (Wood-SEE-ah ill-VEN-sis) *Polypodiaceae*
Woodsia is from the name of the English botanist and architect Joseph Woods (1776-1864), who studied roses. *Ilvensis* means of, or pertaining to Elba, the island off the coast of Italy where Napoleon spent his first exile (1814-1815).

Rusty woodsia is common in dry rocky areas on Precambrian or acid rocks throughout the boreal area. This tenacious little fern grows in dense tufts in soil pockets on the rocks. Easy to overlook because of its size, rusty woodsia is held tightly in place by its stout rhizome. Even treading upon it does not seem to disturb it. This fern has a much denser appearance than the taller and airier-looking fragrant shield fern (*Dryopteris fragrans*, p. 3).

Fronds

10-20 cm high; pinnae oblong, sometimes toothed or turning under at edges; rusty-coloured chaff covers underside.

Field Notes:

Woodland-horsetail

Woodland-horsetail **Horsetail Family**
Equisetum sylvaticum (*Eck-wee-SEE-tum sil-va-TEE-cum*) *Equisetaceae*
Equisetum is derived from the Latin *equus*, meaning "horse," and "*seta*," meaning bristle. *Sylvaticum* means "of the woods."

The **woodland-horsetail** grows in muskeg forest and some sandy areas across the North almost to the treeline. It is a short spiky plant that makes a dramatic show in the spring when it is in its reproductive phase. Growing about 15 cm high, the hollow, jointed stems are topped by a cone-like structure called a strobilus, which contains the spores by which the plant reproduces. At maturity, the plant resembles a horse's tail.

Leaves

Forming later in season, after reproductive phase, when vegetative phase puts out whorls of fringy green leaves from stem joints.

Plant Notes

Horsetails are ancient plants that grew as tall as trees 300 million years ago. The carbonized remains of these ancient plants are burned today as coal.

In some areas, the plant is known as "scouring rush" because the gritty stems can be used for scrubbing. The grit is actually grains of fine silica, which are abrasive enough to polish silver.

Field Notes:

Wild Plants with Seeds but without Flowers

Black Spruce

Black Spruce **Pine Family**
Picea mariana *(pi-SEE-a ma-ree-AH-na)* *Pinaceae*

Picea comes from the Latin *pix*, meaning "pitch;" *mariana* means "of Maryland," referring to the fact that the black spruce grows only in boreal North America.

Black spruce is one of the most common trees of NWT. The densely packed branches at the crown and the drooping upturned lateral branches make this small, (up to 10 m) slender tree easy to distinguish from other conifers. The bark of its trunk is rough and scaly; the bark of the young twigs has fine hairs on the surface. Black spruce will tolerate poor growing conditions such as wetlands, poorly drained areas, rocky soil and cold temperatures. The slow growth of black spruce produces a dense wood.

Leaves

Needles short, four sided; sticking out on all sides of branches, with many pointing upward.

Cones

Female (seed) cones purplish in colour, eventually turning to dark brown; male cones dark red.

Plant Notes

Black spruce reproduces by seed as well as through a process known as "layering," by which the drooping branches near the bottom of the tree take root in the soil and give rise to new spruces.

The Dogrib of Wha Ti use all parts of the spruce for medicine. For example, they boil the boughs to create a steam that is good for congestion, bad colds and headaches. "Spruce beer" can be made by boiling tree branches, while tea is made by boiling leaves and young stems.

Field Notes:

Creeping Juniper

Creeping Juniper **Cypress Family**
Juniperus horizontalis (*jew-NI-pe-rus ho-ri-ZON-tah-lis*) *Cupressaceae*
Juniperus is the Latin name; *horizontalis* refers to the plant's creeping stems.

Creeping juniper grows in rocky and gravelly places, quite often side by side with ground juniper (*Juniperus communis*, p. 10). The blue-green leaves of creeping juniper grow on long, prostrate stems up to 5 m in length. Creeping juniper clings much closer to the ground than ground juniper, rooting freely as it goes, forming mats. It looks somewhat like a prostrate cedar.

Leaves

Soft to the touch, scale like; overlapping.

Flowers

Female cones blue, berry like.

Plant Notes

The Dogrib use tea made from creeping juniper for colds.

Field Notes:

Ground Juniper

Ground Juniper — **Cypress Family**
Juniperus communis (*jew-NI-pe-rus kom-MEW-nis*) — *Cupressaceae*
Juniperus is the Latin name; *communis* means common.

Ground juniper is a common sight across the North, growing to the tree line and slightly beyond. Its spreading branches and bright blue berries are seen wherever there are woods and barren places. Ground juniper grows as a prostrate, spreading shrub. Plants may be male or female. Female cones are known as "juniper berries."

Leaves

Sharp, needle like; dark green below; striped with white line above.

Cones

Male cones small, appearing near ends of branches; young female cones whitish-green, turning bluish-purple in second year of growth.

Plant Notes

Especially popular in Scandinavia, juniper berries can be used to flavour food. A delicious and sprightly marinade for wild meat can be made from the berries. In Britain and France, the berries are used as a pepper substitute and as a base for making beer.

Every part of the juniper shrub can be used for medicine. From treatment for urinary and bladder infections to use in anti-inflammatory compresses. Juniper plants have been used for centuries by native peoples and by old country herbalists. The first written record of juniper's healing properties comes from a 1550 B.C. Egyptian papyrus, which recommends juniper berries as a treatment for tapeworm.

Field Notes:

Jack Pine

Jack Pine | **Pine Family**

Pinus banksiana (PEE-nus BANKS-ee-an-a) | *Pinaceae*

Pinus is the Latin name for "pine," *banksiana* is from the name of Sir Joseph Banks (1743-1820), a British explorer and naturalist.

Jack pine is common in the North, tolerating thin soils and rocky areas. Trees growing in the open tend to be twisted and straggly; those growing in stands tend to be taller and straighter, reaching heights of over 15 m.

Leaves

Yellow-green; sharply tipped; needles in pairs.

Flowers

Male cones small, growing at tips of branches; releasing clouds of pollen when touched; female cones larger, growing farther down branches; tan coloured, hard and slightly curved.

Plant Notes

The young cones of jack pine are edible. The inner bark can be eaten raw or cooked, and the needles can be used to make a tea high in Vitamin C. Squirrels dine on the seeds of jack pine, which are released from the cones after three years. Pine oil is used as an expectorant in cough medicines as well as being used in disinfectants, insecticides, antiseptics, and as a parasitical agent. The amber-coloured resin of jack pine was once used by native North Americans as a treatment for pneumonia, rheumatism and muscle aches and pains.

Field Notes:

Tamarack

Tamarack **Pine Family**

Larix laricina *(LA-riks LA-ri-seen-a)* *Pinaceae*

Larix is the Latin name for larch; *laricina* means "larch-like."

Tamarack has recently replaced jack pine as the territorial tree for NWT. Tamarack is the only member of this genus that grows in this region. Reaching a height of 5-6 m, tamarack has scaly bark and long, slender branches with small woody protrusions that carry the bundles of needle-like deciduous leaves. Tamarack, like black spruce, can tolerate wet and boggy areas and grows just to the tree line in boreal North America.

Leaves

Clustered in groups of 10-20 laterally on branches; soft green colour that turns yellow in autumn before needles drop from branches.

Cones

Male and female cones grow on same tree; female cones up to 2-5 cm long, brown and erect; shed the following year.

Plant Notes

The Dogrib used tamarack as a good, all-around remedy for a number of ailments including stomach problems, colds and fevers, flu and pain relief.

The South Slavey boiled the fresh roots and used the liquid on cuts to speed healing.

The Sahtu Dene use tamarack for treatment of diabetes and arthritis. They believe that the higher the elevation at which the the tree grows, the more powerful the medicine.

Field Notes:

Wild Plants
with both Seeds and Flowers

Black Currant

Black Currant **Saxifrage Family**
Ribes hudsonianum (ry-BEES Hud-so-NEE-ah-num) *Saxifragaceae*
Ribes comes from the Arabic or Persian word *ribas*, which means acid tasting, referring to the berries. *Hudsonianum* refers to "Hudson Bay," which accounts for another common name for this plant, Hudson Bay Currant.

Black currant is a fairly short (1-2 m high) rangy shrub with grey bark. The leaves are shaped like maple leaves with three lobes. The inflorescence is racemose; flowers are replaced by smooth, black berries later in the season.

Leaves

Three lobed; hairy; glandular; edges toothed.

Flowers

8-10; small, white, five petalled.

Plant Notes

Black currants are delicious in jams, jellies and desserts. A black currant syrup, which is popular in Britain, is available here in the grocery store. It has a delicious, refreshing flavour that seems like it's bursting with Vitamin C. Poured over cracked ice and mixed with carbonated water, it quenches the thirst on a hot summer's day.

The South Slavey gathered stems, leaves and flowers of black currant to make a remedy for coughs. A decoction of the boiled plants was taken warm or cold. The stems were also dug from under the snow in winter if cough medicine was needed.

Field Notes:

Blunt-leaved Sandwort

Blunt-leaved Sandwort **Pink Family**
Moehringia lateriflora (*Mo-RIN-gee-ah LAT-uh-ree-flora*) *Caryophyllaceae*
Sandwort is also known by the Latin name *Arenaria lateriflora*. *Arenaria* means "of a sandy place;" *lateriflora* means "flowering from or at the side."

Another delicate member of the pink family, **sandwort** grows in willow thickets and open woodlands north to the tree line. The thin stems are simple or branching to about 25 cm and are covered in tiny hairs. Each stem holds aloft a single, white bloom, which stands out from the side of the stem.

Leaves

Stalkless, narrow; elongated egg shaped.

Flowers

White, five petalled, longer than sepals.

Plant Notes

I found this plant at the base of a garbage can in North Arm Territorial Park just south of Rae. Although considered to be fairly common, it is not a species that attracts attention, so can be easily overlooked.

Field Notes:

Bunchberry

Bunchberry — **Dogwood Family**

Cornus canadensis *(kor-NUS kan-a-DEN-sis)* — *Cornaceae*

Cornus is from the Latin word for "horn," and is thought to refer to a robust European species; *canadensis* means "of Canada."

Because of its creeping rootstock, **bunchberry** forms a carpet at the foot of trees and along the forest floor in moist areas and clearings. Its near-perfect symmetry in flower and foliage, as well as the sharp colour contrasts between bloom and leaf, and berry and leaf makes it seem perfect at any time of the season. The short stem (8-15 cm) is hidden by the whorl of bright green leaves.

Leaves

Conspicuously veined; lower leaves small, upper leaves 3-6 cm long.

Flowers

4 white bracts easily mistaken for petals; flowers tightly cupped within bracts; flowers tiny, purple; 4 petals, 4 stamens.

Fruit

Tightly bunched, bright red berries.

Plant Notes

Bunchberries are considered edible, but not particularly palatable. The seeds are hard like poppy seeds and need to be well chewed.

The Gitskan of Alaska use "clusterberries" as a thickener for other berries, which are sun-dried, rolled and cut into cakes for future use.

The Haida sometimes steamed the berries and preserved them in water and in the oil from the eulachon fish.

The berries are known to be food for red squirrels, bears and several bird species.

Field Notes:

Cloudberry

Cloudberry **Rose Family**
Rubus chamaemorus (roo-BUS kay-MAY-mor-us) *Rosaceae*
Rubus is the Latin name for blackberry; *chaemae* is Greek for "on the ground;" *morus* is Latin for "mulberry."

Cloudberry grows close to the ground in moist areas of peat and muskeg throughout NWT. It is one of the earliest plants to bloom here, putting forth its creamy white blossoms in May. Cloudberry sends out runners underground, which root firmly in the peaty soil.

Leaves

1-3; leathery; kidney shaped; toothed at edges; usually in five parts.

Flowers

Rose shaped; five petalled.

Fruit

Raspberry shaped; bright red at first, turning to amber, then pale yellow when ripe.

Plant Notes

Cloudberries are high in vitamin C. The berries make delicious jams and pies, and are highly prized in Scandinavia for making liqueur. The Inuit also favoured cloudberries and preserved them in seal oil. The flowers can be used in salads as an edible garnish, and the fresh or dried leaves can be used for tea. Wilted leaves, however, are best avoided, because they may be slightly toxic.

The Dogrib boil the flower of cloudberry and apply the tea to sore eyes. They also chew the berries and place them on sores and wounds.

Field Notes:

Common Yarrow

Common Yarrow **Sunflower Family**
Achillea millefolium (*a-KIL-lee-a mil-le-FOAL-ee-um*) *Asteraceae*
Achillea is named for the famous Greek hero, Achilles; *millefolium* means "thousand leaves."

Growing up to 40 cm high on a woolly stalk, **common yarrow** is topped by large, flattened flower clusters. Yarrow can be seen growing in weedy disarray along the roads and in waste areas in August. It is strongly aromatic with a spicy, almost medicinal fragrance.

Leaves

Alternate, numerous, feathery; sub-divided into many linear sections.

Flowers

In dense clusters or "heads" (as in sunflowers); ray florets white or pink; disk florets yellow.

Plant Notes

Yarrow is an effective deterrent to mosquitoes and black flies. Rubbing the plant on your clothes and skin, or throwing a few stalks in your campfire will keep these pesky bugs at a distance.

Achilles reputedly used poultices of crushed yarrow to staunch the wounds of his soldiers after battle. Yarrow was also used in incantations by witches as late as the 17th century. The I Ching, the ancient Chinese system of fortune telling, at one time used dried yarrow stems in divination, instead of the coins used today.

Field Notes:

Cotton Grass

Cotton Grass — **Sedge Family**

Eriophorum angustifolium *(ee-ri-o-PHOR-um an-gus-ti-FOAl-ee-um)* — *Cyperaceae*

Erion means "wool" and *phoros* means "bearing;" *angustifolium* means "narrow leaved."

In some areas, **cotton grasses** grow in great numbers, forming pure stands. The fluffy heads of cotton grass nod on a 20-40 cm stalk. Throughout the low-arctic, this sedge can be seen lining bogs, shallow ponds and lake shores. Circular in cross-section, the stem is unjointed, and the characteristic stem ridges of the sedge family are apparent. All perianth parts of this plant are modified into bristles, which are white.

Leaves

2 or more; sheathing; linear.

Flowers

White, several; nodding; spikes of silky strands.

Fruit

Seeds small, three sided, black; surrounded by white "fluff," that carries seeds in wind as plants mature.

Plant Notes

Over 400 years ago, cotton grass was used as a medicine in Northern Europe, but usage has since been discontinued because the side effects were sometimes worse than the actual malady.

Cotton grass is used by the Inuit as wicks for their stone lamps (*qulliq*). The downy heads of cotton grass have also been used to stuff pillows and mattresses and can be used for tinder.

Many other species of cotton grass grow in the North. Another species of the boreal forest is thin-leaved cotton grass, *E. viridi-carinatum*, which is a smaller plant, with smaller flower spikes.

Field Notes:

Grass-of-Parnassus

Grass-of-Parnassus **Saxifrage Family**
Parnassia palustris (par-NAS-ee-a pah-LUS-tris) *Saxifragaceae*
Parnassia refers to Mt. Parnassus in Greece; *palustris* means "of the marshes."

This delicate, mid-summer beauty is not a grass at all, but a member of the large and far-ranging saxifrage family. **Grass-of-Parnassus** thrives along the edges of streams and lakes, where it grows 35-50 cm high.

Leaves

Basal, heart shaped; 1 small, clasping leaf on each stem.

Flowers

Solitary; 5 green-veined white petals, 5 sepals.

Plant Notes

The conspicuous, swollen ovary of grass-of - Parnassus is surrounded by five fertile and five sterile stamens. The sterile stamens (staminodia) are green and divided into 7-15 segments tipped with small, gland-like structures. Insects are attracted to the structures, which look like drops of nectar, and thus do their job of pollination.

Field Notes:

Labrador Tea

Labrador Tea **Heath Family**
Ledum groenlandicum *(LEE-dum gro-en-LAND-i-kum)* *Ericaceae*
Ledum is Greek for the name of a resin-bearing shrub from Cyprus; *groenlandicum* means "of Greenland."

Labrador tea is an erect, 30-80 cm showy shrub that looks somewhat like a rhododendron. It is found in peaty soils all across our area as far as the tree line. Each flower grows from its own slender stalk at the top of the plant. Crushing the leaves releases a dry, woodsy and highly aromatic scent. The edges of the small leaves roll under and the undersides are woolly, which help the plant retain moisture and protect it from the harsh climate.

Plant Notes

Labrador tea was used as a tea substitute or mixed with black tea by early explorers. A light brew of the tea is refreshing, but prolonged steeping or boiling extracts andromedotoxin, which can cause nausea, intestinal cramps, headaches and even death.

Aboriginal peoples used Labrador tea for many disorders including heartburn, headaches, colds and arthritis. The stems of the shrub were used by the Inuit of Baffin Island as a chewing tobacco; and the leaves were used in a smoking mixture. Labrador tea is known in Europe as "moth herb" because it was hung in closets to repel moths, as well as ghosts and illness.

Leaves

Alternate; oblong; evergreen; edges roll under; undersides rusty woolly.

Flowers

White, small; five petalled; pistil, stamens white and protruding.

Field Notes:

Ladies′-tresses

Ladies'-tresses **Orchid Family**
Spiranthes romanzoffiana (Spee-RAN-thees ro-man-ZOFF-ee-an-ah) *Orchidaceae*
Spiranthes is Greek for "flower coil," referring to the twisted inflorescence; *romanzoffiana* is from the name of a Russian patron of science, Count Nikolai Romanzoff, who sponsored an expedition to the Pacific Coast in 1817.

A fairly inconspicuous member of the orchid family, **ladies'-tresses** grows in wet areas in meadows, fens or along beaches. The 10-20 cm stem holds a spiral flower spike that is dense with white or greenish flowers. Ladies'-tresses have a rich and fragrant perfume.

Leaves

Alternate, narrow; larger at base of plant, becoming sepal-like and clasping toward spike.

Flowers

Spike twisted; lower lip white and bent downward.

Plant Notes

I found this plant quite by accident when scouting a sand beach for rushes. It was growing back a little from water's edge, amongst grass-of-Parnassus and small willows at McNiven Beach in Yellowknife.

Field Notes:

Large-flowered Wintergreen

Large-flowered Wintergreen **Wintergreen Family**
Pyrola grandiflora (PI-ro-la GRAND-ee-flow-rah) *Pyrolaceae*
Pyrola comes from a word meaning "pear tree," referring to the shape of the leaves; *grandiflora* means "large flowered."

Large-flowered wintergreen blooms in June in open woodlands and on sheltered, sunny, tundra slopes. Its large white flowers (1.5-2.0 cm across) distinguish it from pink wintergreen, with which it can be confused if not in bloom. True to its common family name, the leaves stay green all winter, but are replaced by new leaves the next spring.

Leaves

Several, basal, long petioled, oval with obvious veins; leaf edges sharply toothed at wide intervals.

Flowers

Scentless; petals creamy white or pinkish on pink stem; anthers yellow.

Plant Notes

My husband found this plant while playing in the outfield at our daughter's T-Ball game. Chasing a ball into the woods at N.J. MacPherson School in Yellowknife, he was amazed to see several examples of this beautiful plant. After the game, he told me of his find. It amazed me that these plants had survived picking or trampling in an area much used by school children. If you are in Yellowknife in late June, a trip to this spot is well worth it.

Field Notes:

Long-stalked Chickweed

Long-stalked Chickweed — **Pink Family**

Stellaria longipes *(stel-LAIR-ee-a long-JEEP-es)* — *Caryophyllaceae*

Stellaria means "star;" *longipes* means "long-stalked."

This **chickweed** grows to about 25 cm in large, loose clusters. It quite often grows in the same area as prickly saxifrage (*Saxifraga tricuspidata*, p. 28), but the orange-spotted petals of the latter set them apart. Long-stalked chickweed is a non-arctic species of dry and disturbed areas.

Leaves

Small, stiff, opposite; becoming sparse toward top of plant.

Flowers

Small, white, star shaped; petals deeply cut, thus appearing to be 10 rather than 5; sepals slightly longer than petals.

Plant Notes

Chickweeds can be used as salad greens or a substitute for spinach. They are high in Vitamin C and minerals. Common chickweed (*S. media*) is the species most often used in this way. Because of its Vitamin C content, common chickweed is good for scurvy. Leaves, stems and flowers of chickweeds are used in botanical remedies. It is used today in creams for rashes and other skin inflammations, although a bath in chickweed tea is also said to relieve skin conditions. Chickweed has been called a magic healer because it will soothe internal inflammations if used externally.

Field Notes:

Lowbush Cranberry

Lowbush Cranberry — **Honeysuckle Family**

Viburnum edule *(vee-BUR-num ed-yew-lay)* — *Caprifoliaceae*

Viburnum comes from *vieo*, which means "to tie." Another species, *V. lantana*, which grows in Europe, is known for the pliability of its branches. *Edule* means "edible."

Lowbush cranberry is an erect shrub, up to 2 m high with smooth, dark grey bark. A wide-ranging plant, it grows from Newfoundland to Alaska in woodland thickets and scrub north almost to the tree line. The distinctive smelling flowers come out early in the season.

Leaves

Opposite; three lobed; edges serrated; 3-5 obvious veins.

Flowers

Small; grow as flat or roundish clusters in leaf axils; petals 5.

Fruit

Juicy red berries; 3-7 in a bunch; each stem originates at same point.

Plant Notes

The berries are high in Vitamin C and were used to cure scurvy at one time. Lowbush cranberry is used by the Tanaina of Alaska for sore throats and for stomach troubles. The South Slavey called this plant "mooseberries," and used the ripe, boiled berries as a cough medicine. The Dogrib used the boiled berries for urinary problems and constipation.

Field Notes:

Northern Bedstraw

Northern Bedstraw **Madder Family**
Galium boreale *(gay-lee-um bo-REE-aalay)* *Rubiaceae*
Galium is from the Greek *gala*, which means "milk." A similar species, *G. verum*, was traditionally used to curdle milk. *Boreale* means "northern."

Northern bedstraw is a pioneering species of dry or gravelly places. A perennial herb growing to a height of 70 cm, bedstraw has stout, sometimes woody stems that are four-angled.

Leaves

Linear; three veined; growing in whorls of 4 along length of stem.

Flowers

Four petalled; creamy white; lightly scented; growing in three-forked clusters toward top of stem.

Plant Notes

Bedstraw was named as such because members of this plant family were traditionally used to stuff mattresses. The coffee plant is also a member of the this family, so it is not surprising that the seeds of bedstraw can be dried, roasted and ground as a coffee substitute, but with the advantage of being caffeine free. Young plants can be cooked as greens and leaves and roots can be used to make teas. Bedstraws were a traditional ingredient in weight loss teas and foods.

The Cree used the roots of northern bedstraw as a red dye for porcupine quills. Similar species have been used in other countries as a natural source of red dye.

Field Notes:

One-flowered Wintergreen

One-flowered Wintergreen **Wintergreen Family**
Moneses uniflora (mon-uh-SEES yu-nee-FLO-rah) *Pyrolaceae*
Moneses is Greek for "single delight;" *uniflora* means "one flowered."

Forested areas with deep, moist moss are the habitat for **one-flowered wintergreen**. The 10 cm stalk is leafless and holds a single, nodding flower, thus accounting for its other common names of Shy Maiden and Single Delight. If you get down on your hands and knees to look at the flower's underside, you will notice that the stamens, ovary and pistil are conspicuous and swollen. After fertilization, the petals fall away, and the stalk straightens and stands erect as the seed capsule develops.

Plant Notes

The delicate fragrance of this wintergreen is similar to lily-of-the-valley; not wintergreen. It is named "wintergreen" because the leaves stay green all year. Oil of wintergreen is actually derived from a totally different plant, False Wintergreen (*Gaultheria procumbens*), which does not grow in this area.

One-flowered wintergreen has mycorrhizae on its roots, meaning that the roots are associated with a fungus. The fungus may cover the root or even penetrate it; thus the root gets its nutrients through the fungus and not through contact with the soil. Because of the symbiotic relationship with the fungus, one-flowered wintergreen does not transplant well.

Leaves

Basal, small, oval, leathery.

Flower

Waxy white; petals 5, triangular.

Field Notes:

Prickly Saxifrage

Prickly Saxifrage — **Saxifrage Family**

Saxifraga tricuspidata (*sax-ee-FRAY-gah tri-cus-PI-da-ta*) — *Saxifragaceae*

Saxifraga is from the Latin *saxum* (a rock) and *frango* (to break) referring to the plant's habit of growing in cracks of rocks; *tricuspidata* means "three-pointed."

There are over 60 species of saxifrage in North America. **Prickly saxifrage** is 6-15 cm tall, grows only in arctic-alpine regions, and ranges widely across the continent.

Leaves

Basal, leathery, closely packed; tipped with three prickly lobes.

Flowers

Tiny, 5 white petals with small orange spots; 5 sepals, 10 stamens.

Plant Notes

One of the earliest plants to bloom in the North, prickly saxifrage is one of those rare species that seem to have found a way to grow in the most extreme conditions. Forming large mats in waste areas and clinging to the soil in rock crevices, this plant's ability to flourish is amazing. There are conflicting opinions on whether saxifrage actually breaks the rock on which it grows, or grows in soil deposits in the crevices.

This species may be one of the early medicinals used for diseases of the kidney and bladder by early North American peoples. It also fits the floral description of the species used by Roman doctors for urinary problems, especially kidney stones.

Field Notes:

Red Baneberry

Red Baneberry **Buttercup Family**
Actaea rubra (*aac-TAY-a roo-BRA*) *Ranunculaceae*
Actaea means "elder tree," referring to the shape of the leaves; *rubra* means "red."

This wide-ranging boreal species grows in moist meadows and sunny forest openings. A large striking plant with woodsy green foliage and cranberry red berries, **red baneberry** looks more like a shrub than a herbaceous wildflower. One to several leafy stems (up to 1 m high) arise from a thick, fleshy rhizome. This is the only species of baneberry that grows in NWT.

Plant Notes

Although this plant is a member of the buttercup family, it does not at all look like the common yellow buttercup that we all know so well. It does, however, have the deeply cut leaves and many-stamened flowers of this family.

The common name of this plant indicates that perhaps the berries may not be good for consumption, and indeed they are not. All parts of baneberry are extremely poisonous. Even though toxic in nature, red baneberry has been combined with other plants by Dene from Saskatchewan to treat bleeding from the nose. Because of the possibility of improper dosage, however, it is not recommended for use.

Leaves

Alternate, numerous; sharply toothed; divided 2-3 times into threes.

Flowers

White, small; long stalked; grow in long clusters at tip of stem.

Fruit

Seedy; red, glossy; clustered in shape of a bottle brush.

Field Notes:

Red Osier Dogwood

Red Osier Dogwood **Dogwood Family**

Cornus stolonifera (KOR-nus stol-low-NI-fe-ra) *Cornaceae*

Cornus is the Latin name for "dogwood;" *stolonifera* refers to the plant's rooting habit.

This 1-3 m tall deciduous shrub with reddish aromatic bark grows in the Mackenzie River Valley in woods, thickets and clearings. **Dogwood** sends out horizontal branches (stolons) that take root and develop new plants at the nodes.

Leaves

Oval to oblong; surface dark green; undersides lighter green.

Flowers

White, open flat-topped clusters at tips of branches; petals very small.

Fruit

White berries.

Plant Notes

The berries of the dogwood, although bitter in taste, were eaten by all the Interior Aboriginal peoples of BC. They were often mixed with sweeter berries or dried in cakes for later mixing. Dogwood berries make up a large part of the diet of robins. Other birds and wildlife also eat the berries.

The Dene of the South Slave used a decoction of the white berries to treat tuberculosis. The Sahtu Dene use dogwood for sexual diseases, rashes and cleansing the body.

Field Notes:

Round-leaved Orchid

Round-leaved Orchid **Orchid Family**
Orchis rotundifolia (*OR-kis ROW-ton-de-foal-ee-ah*) *Orchidaceae*
Orchis is the Greek name for "testicle;" *rotundifolia* means "round-leaved."

Round-leaved orchid is a delicate, headily scented plant that grows in spruce bogs and moist areas in open woods. Easy to overlook because of its size (10-20 cm), round-leaved orchid is a thrilling find because of its beauty and scent.

Leaves

Solitary, basal, round.

Flowers

2-8; whitish, slightly spaced on both sides of spike; lip purple-spotted.

Plant Notes

At one time, the fresh, fleshy roots of orchids were used in potions that were administered by witches for their aphrodisiac properties. Orchids are prized the world over, and are a protected species in many areas. Plant lovers and trail walkers are encouraged to appreciate them without picking them.

Field Notes:

Saskatoon Berry

Saskatoon Berry **Rose Family**
Amelanchier alnifolia (a-me-LAN-kee-er al-ni-FOAL-ee-ah) *Rosaceae*
Amelanchier is taken from the name of a similar species that grows in Provence; *alnifolia* means "alder leaved."

Saskatoon berry grows up to 2 m in open woods where it may form thickets or colonies. The bark is smooth, dark grey or brown.

Leaves

Alternate, oval, rounded at tip; edges toothed, especially toward tip.

Flowers

Axillary; white, spoon-shaped, separated by 5 pointed green sepals.

Fruit

Large, dark blue berries.

Plant Notes

Saskatoons are one of the most popular berry shrubs in the North. They are delicious in pies, muffins and jams. Aboriginal peoples have used them fresh, or dried and mixed with dried, powdered meat to make pemmican. Aboriginal peoples of the Canadian prairie provinces have used roots and stems to treat lung problems, muscle spasms and pinched nerves.

Field Notes:

Water Parsnip

Water Parsnip **Carrot Family**
Sium suave (*SEE-um swa-VAY*) *Apiaceae*
Sium is from the ancient Greek name of this plant; *suave* means "sweet."

Water parsnip is a tall (0.5-1.0 m), widespread water-loving plant with smooth, hollow stems. It inhabits wet meadows and muddy areas along streams and lakes of the upper Mackenzie drainage area.

Leaves

Opposite; pinnate; long, linear, toothed; glossy-green; leaflets many, 5-10 cm long.

Flowers

In an umbel; small, greenish-white, faintly fragrant.

Plant Notes

Water parsnip may be confused with the poisonous water hemlock (*Cicuta* spp.) because they look so much alike. If the plant has an involucre of narrow segments and a fibrous root, it is water parsnip; if the plant does not have an involucre, and its root is a stout, branched rhizome, it is water hemlock.

Raw or steamed water parsnip roots were used as food by Aboriginal peoples of the British Columbia interior, but it is recommended that readers do not eat any part of this plant in case of an error in identification.

Field Notes:

Water Plantain

Water Plantain **Water Plantain Family**
Alisma triviale (*a-LIS-ma TRIV-ee-el*) *Alismataceae*
Alisma comes from a Celtic word meaning "water;" *triviale* means "ordinary."

Water plantain shows its veined, bright green leaves well above the water line in the southwestern region of NWT, where it grows from a fibrous root anchored in the bed of sloughs and wet areas.

Leaves

Oval; long stalked; growing from base of plant; 5-15 cm long.

Flowers

Small, white or pinkish; in a panicle.

Plant Notes

I found this plant on a hot mid-summer's afternoon while driving from Fort Resolution to Hay River. Sloughs abound along this road, and many of them have water plantain growing in and around the edges. There were several plants at the place we stopped. Earlier in the day, when the light was different, we had not seen the plants as we drove by, but the sun seemed to shine directly on them in the late afternoon.

The Woods Cree of Saskatchewan used the stem base of water plantain as a medicine. It was mixed with water and used as a treatment for heart problems, indigestion and to prevent fainting during childbirth.

Field Notes:

Water-arum

Water-arum | **Arum Family**
Calla palustris (*KA-la pa-LUS-tris*) | *Araceae*

"*Calla*" is from the Greek *kallos*, which means "beautiful." *Palustris* means "growing in bogs."

Water-arum is one of two members of the arum family that grow in NWT, in wet and boggy areas. It can be found at edges of ponds and lakes, and often extends into the water as large mats. The flower is distinguished by its short, cylindrical spadix (1.5-2.5 cm) and its creamy white spathe (2.5-7.0 cm), which looks like a large solitary petal.

Leaves

Oval to heart shaped; bright green; held above water on long stalks attached to a fleshy rhizome embedded in mud.

Flowers

Tiny; yellowish; tightly packed on spadix.

Fruit

Red; berry-like.

Plant Notes

The leaves and roots of water-arum can cause intense burning of the mouth. The South Slavey, however, used fresh or dry rhizomes to soothe a sore mouth. Only the juice of the root was swallowed, not the pulp.

Field Notes:

White Camas

White Camas **Lily Family**

Zigadenus elegans (Zig-ah-DEE-nus elly-GANS) *Liliaceae*

Zigadenus is from the Greek for "joined gland," referring to the two lobes of the heart-shaped nectary; *elegans* means "elegant."

White camas is a deceptively benign-looking plant. It grows in gravelly areas, sunny spots and on river banks early in the season in the upper Mackenzie River valley and mountains, north to the delta. The unbranched stem grows from a bulb to about 60 cm. The leaves arise near the base and few-to-several white flowers grow in a loose cluster at the tip of the stem.

Leaves

Pale green; long and thin like grass.

Flowers

Perfectly symmetrical, perianth of 3 white petals and 3 white sepals; distinctive greenish nectary toward base of petals.

Plant Notes

All parts of white camas are extremely poisonous. It contains the toxic alkaloid zygadenine. Because of the similarity of the bulb, this plant can be confused with wild onion. Wild onion has a distinctive onion smell; white camas has a somewhat foul odour. If in doubt, avoid completely. Two bulbs may be fatal. Even so, some Aboriginal groups have used small amounts of white camas for heart problems. The plant's poison will lower blood pressure and slow down the heart, but misjudging the dose can be fatal.

Field Notes:

Wild Chamomile

Wild Chamomile **Sunflower Family**
Matricaria maritima (*ma-tri-KARE-ee-ah mar-i-TY-mah*) *Asteraceae*
Matricaria means "dear mother," referring to the plant's great medicinal value. *Maritima* means "of the sea."

This cosmopolitan weed will be familiar to anyone who has ever traversed meadows or walked along the roadside. Growing to 50 cm or more, **wild chamomile** has grooved stems that may branch toward the top of the plant. Each branch carries a flower at its tip. This plant is the seaside species that has adapted to drier habitats.

Leaves

Bright green; deeply and finely divided; appear fringe-like.

Flowers

One to several; daisy-like; rays pure white; disks dark gold.

Plant Notes

Species of chamomile have been used since ancient times for their healing properties. Contemporary herbalists use it externally for wound healing and to treat inflammations. Internally, it is used for fever, indigestion, anxiety and insomnia.

In the well known children's story, Peter Rabbit's mother made chamomile tea for him when he returned from Farmer MacGregor's garden. We can assume that he was suffering from not only indigestion, but also anxiety!

Chamomile was one of the plants used in Medieval times to make turf seats. During the days of the Tudors chamomile was planted for fast growing, aromatic lawns. A lawn of chamomile still grows on the grounds of Buckingham Palace.

Field Notes:

Wild Strawberry

Wild Strawberry **Rose Family**
Fragaria virginiana (fra-GAH-ree-ah vir-GIN-ee-ann-ah) *Rosaceae*
Fragaria means "fragrant;" *virginiana* means "of Virginia."

Wild strawberries are not as common in this area of the country as elsewhere. This species is the common strawberry of the interior Northwest, but does not extend far beyond the Mackenzie Valley. The rootstocks send out runners along the ground, producing roots and new plants at the nodes.

Leaves

Arise from base of plant on long stalks; divided into 3 leaflets with toothed edges.

Flowers

5 white petals, 5 sepals, numerous stamens.

Plant Notes

Strawberries have been enjoyed as one of the sweetest fruits of summer for centuries. Raw or cooked, frozen, dried, made into pies, jams and jellies, strawberries are worth the search and the picking time.

A pleasant tasting tea can also be made from the leaves. It is recommended for pregnant women and nursing mothers. It has also been used as a wash for skin irritations, as a gargle for oral irritations and as a stain remover for teeth.

Field Notes:

Arctic Poppy

Arctic Poppy **Poppy Family**
Papaver spp. *(pa-PAH-ver)* *Papaveraceae*
Papaver is the Latin name for this plant.

As its name implies, **Arctic poppy** grows in Arctic or high-alpine areas. The large, creamy or yellow petals are shaped like saucers to catch the sun. This helps the seeds develop faster. The flower heads rotate to follow the sun in its path across the northern summer sky.

Leaves

In basal rosette; pinnately divided; sparsely hairy.

Flowers

Large, showy; may be white, yellow or orange; 4 petals, 2 sepals.

Fruit

Seed capsule; barrel shaped; covered in dark hairs.

Plant Notes

The narcotic juice from poppies was traditionally used to induce sleep and relieve pain. In ancient times, sleep was considered the great physician and the great consoler of human nature. Because of this, the god of sleep was adorned with a wreath of poppies.

Red poppies are worn on Remembrance Day, November 11th, to commemorate the end of the Great War (1914-1918).

Field Notes:

Arnica

Arnica **Sunflower Family**
Arnica amplexicaulis (AR-ni-ka am-PLEX-i-call-us) *Asteraceae*
Arnica means "lamb's skin," referring to the woolly bracts. *Amplexicaulis* means "stem clasping," referring to the leaves.

This species of **arnica** loves moist woods in subalpine areas of North America. Growing 50-60 cm tall, this plant is a hybrid, with variable leaves on the same plant.

Leaves

Opposite; lance-shaped; sharply toothed at edges.

Flowers

Usually 3-5; ray florets bright yellow; disk florets slightly darker.

Plant Notes

Arnicas are difficult to identify because this family comprises many genera of yellow-petalled flowers, and because the arnica genus readily hybridizes. Knowing that arnicas have opposite leaves, rather than alternate, will help you rule out some of the other very similar looking genera, such as groundsels (*Senecio* spp.).

All species of arnica are poisonous. They should never be taken internally, although a tea made from arnica is good as a rub for swollen or bruised feet and legs, but should never be used on broken skin. Creams and lotions made from arnica are well known in Europe, where they are used for tired and aching legs. Such preparations are now available in Canada in stores that specialize in natural products.

Field Notes:

Beggar-ticks

Beggar-ticks **Sunflower Family**

Bidens cernua (*BY-dens sir-NEW-ah*) *Asteraceae*

Bidens means "two teeth," referring to the barbed teeth at the tips of the fruits; *cernua* means "drooping," referring the flower's habit of nodding downward when in fruit.

Growing up to 40 cm, **beggar-ticks** can be found along stream banks, moist meadows and at the edges of ponds throughout the circumpolar world. It is the distance between the ray flowers that distinguishes this plant. The petals are not tightly packed, but separated by an obvious space. The barbs on the fruits are the method of seed dispersal. The fruits readily attach themselves to clothing or animals, which also accounts for the common name of beggar-ticks, as if the seeds are begging a ride.

Leaves

Opposite; leaf stalk or petiole absent; surfaces smooth; edges toothed or smooth.

Flowers

Central disk; ray florets 6-8.

Plant Notes

I found this plant at Niven Lake in Yellowknife. It was growing, in places, in company with duckweed, which had attached itself to the base of the freely branching stems. One of the specimens I collected now sits in my husband's office, dried, pressed and framed. The combination of the sunny face of the flower and the delicate strands of duckweed bring a bit of nature to his office.

Field Notes:

Canada Goldenrod

Canada Goldenrod **Sunflower Family**
Solidago canadensis (so-li-DAH-go kan-a-DEN-sis) *Asteraceae*
Solidago is from the Latin for "to make whole," referring to the plant's medicinal properties; *canadensis* means "of Canada."

Canada goldenrod is more commonly recognized than some of the other species of goldenrod that grow in our area. The leafy stems of this plant grow up to 1.3 m from a branched rhizome. It is common all along the upper Mackenzie River as well as in the northern parts of the eastern provinces and west to the Rockies, Yukon and central Alaska.

Leaves

Lance shaped; edges serrated.

Flowers

In branched panicle; several 5-6 mm flower heads, sometimes heavy enough to bend stem.

Plant Notes

The pollen of Canada goldenrod is heavy and therefore not carried very far by the wind. It is now believed that goldenrod does not cause hay fever, but the plants with which it grows (such as ragweed) are the real culprits.

Canada goldenrod is one of the goldenrod species that is specifically recommended for medicinal use. For on-the-trail first aid for cuts, insect bites and stings, crush the leaves and flowers and apply to the afflicted area.

An American Indian group, the Meskwakis, believed that some children were born without the ability to laugh. To ensure cheerfulness, the people would boil goldenrod with the bone of an animal that died when the child was born. The resulting liquid was used to bathe the newborn in the hopes of ensuring a sunny disposition.

Field Notes:

Common Dandelion

Common Dandelion **Sunflower Family**

Taraxacum officinale (*ta-RAKS-a-cum of-FEE-see-na-lay*) *Asteraceae*

Loosely translated, *Taraxacum* means "disorder remedy," referring to the plant's long history of medical use. *Officinale* means "medicinal."

Dandelions are common in fields, roadsides, lawns and gardens, but grow best in moist, grassy places and recently disturbed ground. Common dandelion is a native of Europe, but has spread all over North America, and most of the world. It is quite often the first flower that very young children reach for, because of its bright yellow colour and ready availability. Dandelions are perennials that grow 40-50 cm high from a long, thick, brown-skinned taproot that tenaciously holds the plants in the soil. All of its parts contain a milky juice.

Leaves

Basal rosette; edges jagged, standing straight up reaching toward top of plant.

Flowers

Heads solitary; ray florets only.

Fruit

Seed heads globe-shaped, fluffy, disperse with slightest puff of breath or wind.

Plant Notes

Dandelion is a versatile food plant, high in iron, calcium, phosphorus, potassium, copper, and vitamins A, B and C. The leaves, when picked young and tender, can be used as raw or cooked greens in a salad. Leaves can even be salted and fermented and used like sauerkraut.

The root can be cleaned, dried, chopped and roasted and then ground into a pleasant coffee. The flowers can be used to make a wine and the whole plant can be brewed as beer.

Field Notes:

Dwarf Goldenrod

Dwarf Goldenrod **Sunflower Family**
Solidago decumbens (so-li-DAH-go de-COME-bens) *Asteraceae*
Solidago means "to make whole," referring to the curative powers of this plant; *decumbens* means "prostrate."

Although goldenrods are widespread in many parts of Canada, **dwarf goldenrod** tends to grow in more northerly regions in open woods or along bodies of water. The dark purple stems of this goldenrod grow to 20-40 cm from a short rhizome. Later in the season the flowers open up and the plant takes on the more characteristic look of a goldenrod.

Leaves

Alternate, narrow; lance shaped.

Flowers

Appear to be single spike of tightly curled blossoms on either side of upper stem (raceme).

Plant Notes

Goldenrod will stop the flow of blood; in some places it is known as "woundwort" because of this characteristic. The leaves of other species have been dried and powdered and used to heal sores. The tea at one time was recommended for stomach disorders. *Solidago* spp. were called sun medicine by the Chippewa and were used to treat fevers, ulcers and boils.

Goldenrod leaves and flowers produce a superior golden-coloured dye.

Field Notes:

Labrador Lousewort

Labrador Lousewort **Figwort Family**
Pedicularis labradorica (*pe-DEE-cu-lair-is lab-ra-DOOR-ee-ka*) *Scrophulariaceae*
Pedicularis is Latin for "louse;" *labradorica* means "of Labrador."

There are many **louseworts** growing in NWT. Several species grow in tundra areas, but this species is common in muskeg and open mossy areas. Growing to a height of 30 cm, the hairy stem is quite branchy, with loose clusters of flowers growing in the leaf axils or at the tip of the stem.

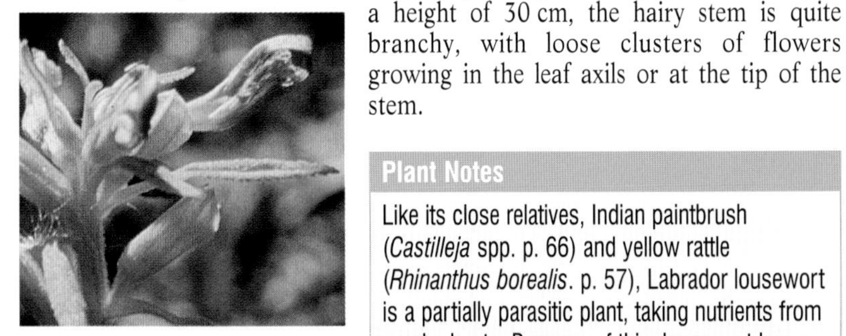

Leaves

Alternate; coarsely toothed.

Flowers

Bilaterally symmetrical; tube three petaled, fused; helmet purplish, two-toothed.

Plant Notes

Like its close relatives, Indian paintbrush (*Castilleja* spp. p. 66) and yellow rattle (*Rhinanthus borealis*. p. 57), Labrador lousewort is a partially parasitic plant, taking nutrients from nearby hosts. Because of this, lousewort has a weak root and the plants can be easily pulled from the ground. They are not, however, easy to transplant because they need a particular host. The flowers of Labrador lousewort are similar, at first look, to the flowers of yellow rattle, and the plant has the same not-quite-in-full-bloom or just-finished-blooming look.

Field Notes:

Marsh Ragwort

Marsh Ragwort **Sunflower Family**

Senecio congestus (se-ne-KEE-oh con-JESS-tus) *Asteraceae*

Senecio means "old man," and *congestus* means "crowded or brought together." It may be assumed that "old man" refers to the dense white fluff of the ripe flower heads, and that "crowded" refers to the dense collection of flowers.

Marsh ragwort, as its name states, is commonly found in wet areas such as marshes. It is a large plant (up to 1 m high) with a thick, slightly hairy stem growing from a short and fibrous root. The stem is hollow and easily compressed. The plant has numerous bright yellow to gold flowers that grow in a fairly compact roundish shape at the top of the stem. The flowers release clouds of white fluff when they go to seed.

Leaves

Alternate; long, linear, clasping, wavy edged.

Flowers

Yellow; disk and ray florets present.

Plant Notes

The young leaves and flowering stems are edible and may be used in salads or cooked. Caution is advised when gathering plants, however, to ensure the water near which they grow is not polluted.

Marsh ragwort is one of over 1300 species of *Senecio* world-wide; of these, 16 species have been recorded in NWT. This species is quite easy to identify because of its habitat and its height. Usually marsh ragwort flowers only once and then dies, but some of the plants growing in our higher latitudes may flower and fruit the following year if the first year's flowers are killed by an early frost.

Field Notes:

Narrow-leaved Hawkweed

Narrow-leaved Hawkweed — **Sunflower Family**

Hieracium umbellatum (*hi-RAY-cee-um um-BEL-lay-tum*) — *Asteraceae*

Hawkweeds get their common and genus name from the Latin *hierax*, which means "hawk." An old English belief is that hawks swooped down to earth to get the juice from the plant to sharpen their eyesight. *Umbellatum* refers to its umbellate flower clusters.

Many large flowers top the erect, 1 m stem of this weedy-looking native of meadows and woodland clearings. **Narrow-leaved hawkweed** is widespread in Canada, from New Brunswick to British Columbia and north to Great Bear Lake. Hawkweeds are a huge genus, with over 800 species worldwide. This plant contains milky juice, like sow thistle (*Sonchus arvensis*, p. 53) and common dandelion (*Taraxacum officinale*, p. 43).

Leaves

Stiff; oblong; larger toward bottom of stem; edges may be smooth or toothed. Leaves on upper part of stems are stalkless and have clasping bases.

Flowers

Similar to common dandelion with ray florets only.

Plant Notes

Folk medicine accorded a good reputation to hawkweed as a treatment for jaundice when the milky juice was added to wine and given morning and night. The bruised green plant was also used in medicine by mixing it with salt to put on wounds. Hawkweeds are reported to soothe the stomach and aid digestion. American Indians of the Northwest used a native species as a chewing gum.

Field Notes:

Rough Cinquefoil

Rough Cinquefoil **Rose Family**
Potentilla norvegica (*po-ten-TIL-la nor-VEG-i-ka*) *Rosaceae*
Potentilla means "potent;" *norvegica* means "Norwegian."

Rough cinquefoil is of weedy habit and grows in disturbed soil, clearings or burns in non-arctic parts of the world. The simple stems are leafy and hairy and can grow up to 60 cm. In some plants, the stems branch toward the top. This species is just one of the many cinquefoils that grow in the North.

Leaves

Deeply cut into 3 coarsely toothed lobes; leaves closer to base of plant long-petioled.

Flowers

Small, petals pale yellow; sepals long, peeking between petals.

Plant Notes

I found this plant on a rock bluff on Walsh Lake, growing right at the edge of a fire pit. At first look, I wasn't sure whether it was a buttercup or a cinquefoil. Both have five-petalled, yellow flowers and deeply toothed leaves. The buttercup, however, does not have stipules at the base of the leaves or the characteristic hypanthium of members of the rose family.

Field Notes:

Shrubby Cinquefoil

Shrubby Cinquefoil **Rose Family**
Potentilla fruticosa (*po-ten-TIL-la froo-ti-KOE-sa*) *Rosaceae*
Potentilla means "potent;" *fruticosa* means "shrubby."

Shrubby cinquefoil thrives in peaty environments throughout the North. A deciduous plant, cinquefoil forms low bushes up to 1.5 m tall. The bark is reddish-brown and tends to shred easily. As a member of the rose family, the flowers look like small, wild roses, growing near the tips of the branches singly or in a small spray. Shrubby cinquefoil first blooms in June and keeps blooming throughout the summer.

Leaves

Smooth edged; divided into 5 leaflets ("cinquefoil"); silky hairy on both sides.

Flowers

Five petalled; bright yellow; stamens numerous.

Plant Notes

This entire plant can be gathered as medicine. It is astringent and said to reduce inflammation of gums and tonsils. One or two cupfuls of a decoction of the stem, leaves and roots were given for fever accompanied by aching limbs by the South Slavey. Because of the plant's high tannin content, Native North American groups used the plant to stop bleeding as well as to treat diarrhea and dysentery. Cinquefoil was also considered to be a magical plant by some Native North American groups. It was thought to provide protection against evil, especially witches.

Field Notes:

Silverweed

Silverweed — **Rose Family**

Potentilla anserina *(po-ten-TIL-la an-se-REEN-ah)* — *Rosaceae*

Potentilla means "potent;" *anserina* is Latin for "goose."

Silverweed is a non-arctic species that can be found along the Mackenzie River and its tributaries. Growing on gravelly or sandy lake shores and beaches, silverweed takes root by means of stolons, which creep along the ground, producing roots and new plants at the nodes.

Leaves

Fringy; silver below, but green above.

Flowers

Petals 5; growing from axils of leaves on newly rooted runners.

Plant Notes

Silverweed has been gathered as food since prehistoric times. Its starchy roots are said to taste like chestnuts, parsnips or sweet potatoes. They can be eaten raw, roasted, boiled or fried. In England, dried and ground roots were once used for flour for baking bread. The leaves are tender and can be cooked as greens.

When the weather turns bad, the flowers of silverweed close halfway, and when it starts raining, the leaves crowd together to form a protective roof over the fragile flowers.

The old German name for silverweed is "goose grass," because geese like to eat it, and silverweed likes to grow where geese have trampled the ground and left their droppings.

Field Notes:

Silverberry

Silverberry **Oleaster Family**
Elaeagnus commutata (*e-lee-AG-nus ko-maw-TAH-ta*) *Elaeagnaceae*
Elaeagnus is Greek and originally applied to a willow, which silverberry closely resembles. *Commutata* means "changed."

Silverberry is one of the two species of oleaster growing in our area. The other is soapberry (*Shepherdia canadensis*, p. 104). The 1-2 m shrub is distinguished by its rusty, scurfy twigs and its sweet, heavy scent in spring. Silverberry grows in abundance on the banks of the Slave, Liard, Hay and Mackenzie Rivers.

Leaves

Lance shaped; silvery, scurfy on both sides.

Flowers

Axillary; yellow on inside, silvery yellow on outside.

Fruit

Silvery, shimmery berries.

Plant Notes

The seeds of silverberry have traditionally been used by the Gwich'in people as decoration. The berries are boiled to remove their flesh and while still soft, the seeds are pierced and threaded. After the seeds dry, they are oiled and are ready to be used for jewellery and clothing. The grooved seeds make an attractive bead.

Field Notes:

Small Bladderwort

Small Bladderwort **Bladderwort Family**
Utricularia minor (*Yu-tree-ku-LAH-ree-ah MINE-er*) *Lentibulariaceae*
Utricularia refers to "bladders;" *minor* means "small." Wort is the old English name for "plant."

Small bladderwort is a rootless aquatic plant, about 20 cm long, growing in shallow water as far north as Great Bear Lake. It is a delicate plant with finely dissected submersed leaves, a long, leafless stem, and three to four small, bright yellow flowers. Attached to the leaf stalks are many small, round bladders.

Leaves

Finely dissected; pale green; growing from underwater stem.

Flowers

Irregular; two lipped, spurred; poking out above water.

Plant Notes

Like its close relative, butterwort (*Pinguicula vulgaris*, p. 76), bladderwort is carnivorous. Instead of sticky leaves like the butterwort, bladderwort has tiny bladder-like traps that are actually specialized leaves. These traps will open at the touch of small invertebrates, like mosquitoes, and close upon the prey in the rush of water that comes into the bladder. Digestive enzymes break down the tiny animals to form nutrients for the plant.

I found this plant quite by accident while collecting a specimen of water plantain (*Alisma triviale*, p. 34) on the road from Fort Resolution to Hay River. My husband caught a glimpse of something yellow across the road, and there it was, in a murky pool just at road's edge.

Field Notes:

Sow Thistle

Sow Thistle **Sunflower Family**
Sonchus arvensis (*SON-chus ar-VEN-sis*) *Asteraceae*
Sonchus is Greek for "hollow," pertaining to the hollow stems; *arvensis* means "of cultivated fields," referring to one of the most common areas where the plant grows.

Sow thistle raises its glorious yellow heads in fields, along roadsides and waste places in the warmer regions of NWT. It looks like a very tall (up to1 m) dandelion, but instead of just one flower per plant, it has several flowering heads. The leafy stems contain a milky juice.

Leaves

Alternate; clasping the stem; with prickly edges that roll backward.

Flowers

Ray flowers only, no central disk.

Plant Notes

Sow thistle has been used as a salad herb since ancient times. According to legend, a dish was given to Theseus before he slew the bull at Marathon in Classical times.

A similar species, common sow thistle, *S. oleraceous*, is a well-known folk remedy in Europe, taking its common name from the fact that sows eat a lot of this plant when suckling, thereby increasing milk production. Among its other virtues, the milk was recommended in a drink for short-windedness, used topically for skin eruptions and irritations, and as a beauty-enhancing skin wash.

Field Notes:

Yellow Lady's Slipper

Yellow Lady's Slipper **Orchid Family**
Cypripedium calceolus *(kip-ree-PEE-dee-um cal-see-OH-lus)* *Orchidaceae*
Cypripedium is from the Greek and translates as "Venus slipper." *Calceolus* means "a small shoe."

Yellow lady's slipper is found in moist woodlands with lime-rich soils. It grows along the Liard River and the upper Mackenzie River, north to Norman Wells, western Great Bear Lake and the Mackenzie Mountains. The distinctive feature of this orchid is its "pouch," which looks like a slipper. The pouch is actually the lip characteristic of the orchid family. Yellow lady's slipper grows 15-30 cm tall on a single stalk from coarse, thick roots. The usually single yellow flower is small and fragrant.

Leaves

Large; egg shaped, sheathing stem.

Flowers

Lip pouch shaped; petals lateral on stem, longer than lip, purplish-brown in colour; petals difficult to distinguish from sepals, but petals narrower and twisted.

Plant Notes

Dene of the South Slave used lady's slipper as a love charm. A single strand of a girl's hair was tied about the stem and carried next to a man's heart.

Lady's slippers are becoming rare because of over picking. Once picked, the whole plant dies because it grows in association with fungi in the soil.

Field Notes:

Yellow Mountain Avens

Yellow Mountain Avens **Rose Family**

Dryas drummondii (*DRY-as drum-MOND-ee-eye*) *Rosaceae*

Mountain-avens takes its name from the Greek tree nymphs, the *dryads*, because some of the species have leaves that look somewhat like tiny oak leaves; *drummondii* refers to the botanist who discovered this species, Thomas Drummond (1780-1835).

Yellow nodding flowers and dark green leaves distinguish this species of **mountain avens**. A basal rosette of leaves gives rise to the 25 cm stem. This species of mountain avens is sub-alpine and non-arctic, growing in lime-rich soils.

Plant Notes

Leaves are waxy and slightly rolled under, with white, woolly undersides. This adaptation helps the plant conserve moisture in the summer and to shed ice in the winter.

A similar species, *Dryas integrifolia*, is the territorial flower for NWT. It has white petals and a yellow centre.

I found yellow mountain avens along the rocky shore of the Trout River at the foot of Samba Deh Falls in the same area as lobelia, gentian and harebells. It was easy to spot, because it was going to seed and displaying its characteristic twisted tuft of long, feathery fruits, shining in the sun. I have also seen it growing in the alluvial flats at Fossil Canyon in Norman Wells, as well as on the tundra.

Leaves

Oval; dark green; deeply wrinkled.

Flowers

Yellow; petals 8-10; cup-shaped.

Field Notes:

Yellow Pond-lily

Yellow Pond-lily **Water-lily Family**
Nuphar variegatum (NEW-far VAR-ee-gay-tum) *Nymphaeaceae*
Nuphar is from a Persian or Arabic word meaning "water-lily." *Variegatum* means "varied," referring to varying colours of sepals.

Yellow pond-lily carpets the surface of ponds in many areas of NWT in mid-summer. It is common throughout wooded lowlands west to the Upper Mackenzie Basin and north to Great Bear Lake. From between green floating leaves bobs the compact yellow flower. Long, slender and flattened stalks are attached to a thick underwater rhizome.

Leaves

Large (10-20cm); heart shaped.

Flowers

6 yellow sepals; petals small, inconspicuous; stamens numerous, yellow.

Plant Notes

The flowers of this plant are unusual because what appears to be six yellow petals are actually six yellow sepals. Three different colours are seen inside these sepals: deep purple at the base, changing to green and then to yellow.

The rhizomes from yellow pond lily can be used as food by animals and people. Beaver, muskrat and moose eat the root. Aboriginal peoples boil or fry the roots after slicing them. Roots can also be used as medicine in a poultice for boils, skin ulcers or infected wounds.

Field Notes:

Yellow Rattle

Yellow Rattle **Figwort Family**
Rhinanthus borealis (*ri-NAN-thus bo-ree-AH-lis*) *Scrophulariaceae*
Rhinanthus is Greek for "nose-flower," referring to the distinctive shape of the corolla, which looks like a beak; *borealis* means "northern."

Yellow rattle is common in many dry, gravelly areas in the North, including the areas along the highway south of Great Slave Lake. The simple stems, which can be slightly branchy, are 30-50 cm tall, and the flowers, with their distinctive inflated calyx, bear only a few yellow petals at a time.

Leaves

Opposite, coarsely toothed, nearly stalkless.

Flowers

Yellow, few; calyx inflated.

Plant Notes

Like the Indian paintbrush (*Castilleja raupii*, p. 66), yellow rattle is partially parasitic and takes its nutrients from nearby plants. In this way, yellow rattle, an annual, takes water and nutrition from nearby perennial grasses. When yellow rattle completely dies back, it turns black. After the petals fall away, the calyx turns into a dried seed container that looks like a flattened balloon. When you shake the plant, you can hear the seeds rattling around inside, which gives rise to the common name of yellow rattle.

Field Notes:

Yellow Sweet Clover

Yellow Sweet Clover **Pea Family**
Melilotus officinalis (MEE-lee-low-tus of-FEES-e-nail-is) *Fabaceae*
Melilotus is derived from the Greek *mel* for "honey" and *lotus* for "lotus flower;" *officinalis* means medicinal.

Yellow sweet clover can be recognized by its tall, yellow flower spikes, which wave gently to and fro in the wind. Aptly named, the fragrance of sweet yellow clover will lift your spirits if you simply walk through a field of it. This sweetly scented plant is a favourite of honey bees, which perform the task of pollination as they take nectar from the plant. Sweet clover can be found along roadsides throughout the North. It is originally a Eurasian species introduced as a forage crop.

Leaves

Compound; leaflets 3.

Flowers

Small, pea-like, arranged alternately in spike at tips of branching stems, which may grow up to 1 m in height.

Plant Notes

The vanilla-like taste of sweet clover comes from a substance called coumarin, which is used to make blood-thinners. The dried plant can be used in teas that relieve headache and soothe nerves. Poultices made from sweet clover can reduce inflammation; salves can heal ulcers and burns. The stems, leaves and flowers produce yellow shades of dye. At one time in Europe, yellow sweet clover was put in pillows to encourage happy dreams.

Field Notes:

Alpine Azalea

Alpine Azalea **Heath Family**
Loiseleuria procumbens (*lwe-ze-LUR-ree-a pro-KUM-bens*) *Ericaceae*
Loiseleuria was named for Jean Louis Auguste Loiseleur-Deslongchamps (1774-1849), a French botanist who was an authority on indigenous medicinal plants and an expert on classification; *procumbens* means "trailing along the ground."

Alpine azalea grows only in arctic-alpine regions. It is a mat-forming, dwarf shrub with leathery leaves and tiny pink flowers. Two other members of the heath family, Labrador tea (*Ledum groenlandicum*, p. 21) and blueberry (*Vaccinium uligonosum*, p. 62), are often found close by.

Leaves

Opposite, oval, smooth; edges rolled.

Flowers

Star shaped, five petalled with pointed tips.

Plant Notes

I have seen alpine azalea only three times in my life. The first time was during a wildflower course I attended in Churchill, Manitoba, many years ago. The second was during a hike through Auyuittuq National Park on Baffin Island in Nunavut a few years later. The third was last summer on the tundra at Daring Lake, northwest of Yellowknife. Each of these times, I have been thrilled to find a plant that has made the adaptations that allow it to survive in harsh growing conditions. If you find yourself in an area where you think this plant maybe present, it would be worth the effort to keep your eyes keenly focused on the ground.

Field Notes:

Arctic Dock

Arctic Dock **Buckwheat Family**
Rumex arcticus (RU-meks ARK-ti-cus) *Polygonaceae*
"*Rumex arcticus*" is literally translated as "arctic dock."

As its name suggests, **arctic dock** is a northern species that grows in arctic-alpine areas. Growing in damp or turfy places, its single stem can reach heights of 100 cm. The root is stout and fleshy. Arctic dock is the most common species of dock in northwestern Canada.

Leaves

Mostly basal; long and narrow; wavy at edges.

Flowers

Panicle of small, lateral flowers; tightly packed along stem.

Plant Notes

Most species of dock are edible. The leaves, which are high in Vitamin A and iron, can be used as potherbs and in salads. The leaves and roots contain tannin, so if you are sensitive to this substance, avoid eating the plant.

My first encounter with dock was on a walking holiday in Cornwall, England, a couple of years ago. As my group was walking through an area of meadow, I got a nettle stuck in my finger. The tour leader picked a dock leaf and told me to rub it on the sore spot. To my amazement, it took the sting and pain away. I can now appreciate why Aboriginal groups used dock in poultices to remove slivers.

Field Notes:

Bearberry

Bearberry **Heath Family**

Arctostaphylos uva-ursi (*ark-toe-STAAH-fil-os YU-va -UR-see*) *Ericaceae*

Arctostaphylos is derived from *arctos*, which means "bear," and *staphyle*, which means "bunch of grapes." *Uva-ursi* means "grape of the bear."

A circumpolar plant, **bearberry** often forms the groundcover in coniferous woods. It also grows on exposed rocks, river banks, eskers and sand plains northward to near the limit of trees. It is a branchy, low-growing shrub with peeling reddish or dark grey bark. The flexible branches root freely as they crawl along the ground. Flower clusters grow at the tips of branches and are replaced by red berries in fall.

Two other species of bearberry grow in our area: *A. alpina* (p. 86) and *A. rubra*. Both have toothed leaf margins. Alpine bearberry (*A. Alpina*) is very obvious in autumn, with its shiny, black berries and dark red, wrinkled leaves.

Leaves

Opposite, evergreen, leathery, spoon shaped.

Flowers

Tiny, delicate; shaped as five-segmented, pink-tipped urns.

Plant Notes

Bearberry is also known as "kinnikinnick," which means "smoking mixture," because of its use as tobacco by Aboriginal peoples. Fur traders commonly blended kinnikinnick with regular tobacco to extend supplies.

Bearberry has also been used as a diuretic and antiseptic for more than 1000 years.

Field Notes:

Bog Blueberry

Bog Blueberry **Heath Family**
Vaccinium uligonosum (*vak-SEEN-ee-um u-lee-go-NOH-sum*) *Ericaceae*
Vaccinium is the Latin name for this genus; *uligonosum* means "of wet or marshy areas."

Bog blueberry is a low, much branched, small shrub of acidic soils in the circumpolar world. In some areas it can grow to 60 cm, but is commonly much shorter and compact.

Its beautifully coloured flowers can be seen dotting the tundra in July, soaking up the sun of the short, but bright, northern summer.

Leaves

Alternate, small, oval; margins entire; dull green above, dusky below, turning bright red in fall.

Flowers

Axillary; rosy pink; usually four lobed.

Fruit

Small, dark blue berries.

Plant Notes

Blueberries are a popular source of fruit in late summer. Pies, desserts, muffins, pancakes, bannock and syrups can all be made from this delicious berry.

Dene peoples use the leaves in a tea.

Peoples of the boreal forest have used blueberries in a root decoction for headaches, as part of an anticancer medicine and for problems encountered during pregnancy and childbirth.

Field Notes:

Bog-rosemary

Bog-rosemary **Heath Family**

Andromeda polifolia (*an-DRO-me-da po-LI-foal-ee-ah*) *Ericaceae*

According to Greek mythology, *Andromeda* was chained to a rock in the sea, but was subsequently rescued by her hero, Perseus. *Polifolia* means white-leaved.

Bog-rosemary grows in muskeg, damp tundra and turfy areas north beyond the tree line throughout the circumpolar world. Like other members of the heath family, it is a small, trailing shrub. Bog-rosemary looks somewhat similar to bog-laurel (*Kalmia polifolia*), but bog-laurel has opposite leaves and wheel-shaped flowers. The foliage also resembles Labrador tea (*Ledum groenlandicum*, p. 21), but Labrador tea is rusty woolly under the leaf; bog rosemary is whitish.

Leaves

Alternate; evergreen; edges roll under; top of leaves deeply grooved; undersides white.

Flowers

In terminal clusters; delicate, urn-shaped.

Plant Notes

Bog-rosemary is considered toxic because it contains a poisonous compound called andromedotoxin, which can cause vomiting and death. It can lower blood pressure, cause breathing difficulties and intestinal upsets.

Bog-rosemary is used by the South Slavey to make a decoction to treat stomach ache that is accompanied by body aches and cough.

Field Notes:

Dwarf Raspberry

Dwarf Raspberry — **Rose Family**
Rubus acaulis *(roo-BUS a-CAW-lis)* — *Rosaceae*
Rubus means "red;" *acaulis* means "stemless."

Dwarf raspberry grows on very short stems (5-10 cm) in moist turf up to and slightly beyond the tree line. Unlike other raspberry plants, this species has no thorns but is tufted. The fruit is a small, delicious, red raspberry. Each of the tiny fleshy globes that make up the berry contain a seed.

Leaves

Few, alternate, deeply divided into three; lobes toothed.

Flowers

Showy, solitary, bright pink, fragrant; petals 5, sepals 5.

Plant Notes

There are at least 50 species of *Rubus* in North America grown as commercial crops or gathered for personal use. The sweet taste of dwarf raspberry is a welcome treat for birds and small animals as well as people. Dwarf raspberry was gathered as food by the Dene of the South Slave in late July and eaten fresh or stored in baskets buried in the ground.

The berries and juice of *Rubus* spp. were also used by many North American Indian groups for stomach troubles. The astringency of the berries was said to control diarrhea and dysentery, even amongst infants and young children. Early settlers in America were said to add the juice to honey and alum to tighten loose teeth and to dissolve tartar on teeth.

Field Notes:

Fireweed

Fireweed **Evening Primrose Family**

Epilobium angustifolium (*ep-ee-LO-bee-um ang-GUSTI-fo-lee-um*) *Onagraceae*

Epi is Greek for "upon;" *lobos* means "pod." This refers to the way the petals surmount the pod-like ovary. *Angustifolium* means "narrow-leaved."

One of the North's most common and showy plants, **fireweed** grows in abundance in backyards, ditches and roadsides. Growing up to 3 m, fireweed can be recognized by its long stalk of showy, bright pink flowers. Fireweed gets its common name from its habit of being one of the first plants to grow in a burned-out area.

Leaves

Alternate; long, narrow; look like willow leaves.

Flowers

Bloom from bottom of stalk up; ovary conspicuous, long, reddish; pistil four lobed, white, drooping.

Fruit

Carried in "fluff," which floats through the air to deposit seeds far and wide.

Plant Notes

Fireweed has the curious distinction of displaying three stages of growth on a single plant: red seed pods ready to burst at the bottom of the flowering spike, open magenta blooms in the middle of the spike, and buds just about to bloom at the tip.

All parts of fireweed are edible. The plant produces a fragrant, dark honey. Noon is peak production time for nectar, and bees swarm the blooms at this time.

The Sahtu Dene combine fireweed with water and bear fat for relief from skin problems.

Field Notes:

Indian Paintbrush

Indian Paintbrush **Figwort Family**

Castilleja raupii *(Cas-tay-LAY-yah raw-OO-pee-i)* *Scrophulariaceae*

Castilleja is from Domingo Casillejo, an 18th century Spanish botanist; *raupii* is from Hugh Raup, an American botanist, who, in 1939, made collections at Fort Simpson and at Britnell Lake in the Mackenzie Mountains.

Indian paintbrush stands 20-40 cm along lake shores, river banks and in drainage areas throughout the North. One-to-several stalks branch out from the base of the plant and each stalk holds a bright magenta flower cluster that can be seen from a distance. Several flowering heads grow on each plant.

Leaves

Alternate, slightly hairy, narrow, 4-5 cm long; tend to twist and curl.

Flowers

Yellow, small, enclosed in showy magenta bracts.

Plant Notes

Paintbrush roots are short and weak because they are partially parasitic on the roots of nearby plants.

Because paintbrushes contain chemicals called alkaloids that may cause liver damage, internal use should be avoided.

Field Notes:

Mountain Cranberry

Mountain Cranberry **Heath Family**

Vaccinium vitis-idaea (*vak-SEEN-ee-um vi-TIS-ee-DAY-ah*) *Ericaceae*

Vaccinium is the Latin name for this plant; *vitis-idaea*, means "vine of Mt. Ida." Mt. Ida is located in Crete.

Mountain cranberry inhabits the boreal forest, where it thrives in acidic and peaty soils. It is circumpolar in range, growing in subarctic and arctic areas. A low, trailing, evergreen shrub, mountain cranberry is often confused with bearberry, with which it grows. Both are found in open woods, but cranberry is also found in drier, open, rocky areas.

Leaves

Small, oval, leathery, shiny; obvious mid-vein; undersides dotted with black, bristly points.

Flowers

Pale pink, small; four petalled, bell shaped.

Fruit

Shiny, dark red berries.

Plant Notes

Cranberries are one of the most popular northern berries for human consumption. They are delicious in cranberry sauce, puddings, breads or in a sweetened, cool drink. The berries overwinter and are a good food source for animals during the colder months.

The Dogrib chewed or boiled the berries for colds and coughs. The Sahtu Dene use the cranberry mainly for kidney and bladder problems.

Cranberries are so acidic that they can bleach out discolourations in the skin.

Field Notes:

Pink Corydalis

Pink Corydalis **Fumitory Family**
Corydalis sempirvirens (ko-RI-da-lis sem-peer-VEER-ens) *Fumariaceae*
Corydalis is the Greek name for "lark": the spur of the flower resembles the hind claw of this bird. *Sempirvirens* means "evergreen," referring to the leaves.

Growing in open areas across the boreal forest, **pink corydalis** can be seen seemingly sprouting from rock bluffs. The tall stem (up to 60 cm) branches close to the top in racemes of 3 to 6 flowers.

Leaves

Finely dissected; grow mostly toward base of plant.

Flowers

Irregular; petals 4, tubular, pink with yellowish tips; inner two petals joined at tip; one outer petal pouch-shaped at base.

Seeds

Shiny; contained in pod-shaped, curved capsule commonly occurring on plants still in bloom.

Plant Notes

Insects must force their way in to reach the nectar and thereby brush up against the anthers, picking up pollen.

Field Notes:

Prickly Wild Rose

Prickly Wild Rose — **Rose Family**

Rosa acicularis (*RO-sa asee-KU-la-ris*) — *Rosaceae*

Rosa is Latin for "rose;" *acicularis* means "needle-like," referring to the long, sharp thorns of this plant.

Prickly wild rose is common throughout the District of Mackenzie in clearings, burns and beside water. It is one of three species of wild rose that grow in the North. The other species have smooth stems or only a few scattered thorns; prickly wild rose is aptly named for its bristly or prickly stems. The plant may be straggling or erect and, like cinquefoil, grows up to 1.5 m tall.

Leaves

Compound with 3-9 leaflets; edges serrated.

Flowers

Showy, fragrant, short lived; pink petals silky soft.

Fruit

Rose hips, green at first, turning red in fall.

Plant Notes

Prickly rose was used by the Dene of the South Slave for bee stings. The petals were chewed and placed upon stings to relieve the pain. The Dogrib used rose hip tea to cure mouth infections, a sore stomach, shortness of breath and coughing. In the Sahtu, the rose petals are boiled and used as a wash for sore eyes and cataracts. The hips are made into a beverage for colds and pneumonia. A liquid made from the boiled leaves is used on rashes or to soothe the skin.

Field Notes:

Twinflower

Twinflower **Honeysuckle Family**
Linnaea borealis (*lin-NAY-a bo-ree-AH-lis*) *Caprifoliaceae*
Linnaea was named in honour of Carolus Linnaeus (1707-1778), the father of botanical nomenclature, because it was one of his favourite plants; *borealis* means "northern."

Twinflower is one of the most delicate beauties of the boreal forest. Two nodding, pink flowers grow from the two branches of a forked stem. Twinflower favours mossy or peaty areas throughout NWT.

Leaves

Oval; blunt tipped with toothed edges.

Flowers

8-15 mm; tubular; sweetly scented.

Plant Notes

Although included in this section on wildflowers, twinflower is actually a shrub, which trails along the ground by means of branching stems. The fruits of twinflower are sticky burrs that are spread by animals and people.

Field Notes:

Arctic Lupine

Arctic Lupine **Pea Family**
Lupinus arcticus (*lu-PEEN-us ark-TI-kus*) *Fabaceae*
Lupinus is from the Latin *lupus*, meaning "wolf;" *arcticus* means "of the arctic."

Arctic lupine is endemic in the Northwest, occurring along roadways and open areas throughout the Mackenzie Delta. This bushy plant has pea-shaped flowers and seed pods.

Leaves
Palmately compound, leaflets 6-8; dark green.
Flowers
Blue to purple; in clusters.
Fruit
Seed pods large, hairy; twisted after opening.

Plant Notes
Arctic lupine is easily transplanted to the home flower garden. It is a hardy plant and produces many seeds each year. Although the seed pods strongly resemble the pods of edible garden peas, they should not be eaten. Seeds of many members of the pea family are poisonous.

Field Notes:

Blue Columbine

Blue Columbine **Buttercup Family**
Aquilegia brevistyla (*aah-kwi-LEE-gee-a brev-i-STY-la*) *Ranunculaceae*
Aquilegia is from the Latin *aquila*, which means "eagle," probably referring to the resemblance of the eagle's talons to the shape of the spurs on the flower. *Brevistyla* means "short-styled," referring to the length of the middle part of the female reproductive structure.

The only columbine growing in NWT, this delicate plant finds its home on sunny, rocky slopes north to the limit of the trees. **Blue columbine** grows from a stout rhizome and stands erect with slender 30-80 cm stems.

Leaves

Alternate; occurring mostly at base of stem; twice divided into 3 parts.

Flowers

Inner petals tube-shaped, creamy white, surrounding numerous yellow stamens; 5 outer purplish sepals, spreading out and back from petals; petals have hooked spurs that protrude from between sepals.

Plant Notes

The nectaries of this plant are difficult to reach because they are at the end of the spurs. Only insects with long tongues, such as butterflies and bees, can reach them.

Field Notes:

Blue-eyed Grass

Blue-eyed Grass **Iris Family**
Sisyrinchium montanum (*si-si-RING-kee-um mon-TAN-um*) *Iridaceae*
Sisyrinchium is from the Greek name for another iris-like plant; *montanum* means "of the mountains."

Blue-eyed grass grows mostly in the Mackenzie lowland region, reaching as far north as Norman Wells. It also thrives in the more tolerant conditions south of Great Slave Lake. This member of the iris family has the striking purple petals and yellow centres of the much-loved wild irises and blue flags that grow in more southerly locales. Blue-eyed grass, however, does look like a grass with its narrow, grass-like leaves growing in a tuft from fibrous roots.

Leaves

Ridged, somewhat flattened; growing to half the length of the 20-30 cm stem.

Flowers

Delicate, growing from between two greenish bracts at tip of stem; 6 violet, ridged petals surround yellow centre; petals notched or with an awn at tips.

Plant Notes

The flower bud opens only when the humidity and sunlight are just right, each flower lasting for only a few days.

Teas made from different parts of the plant have been used by Aboriginal peoples for various ailments associated with the digestive system, such as stomach aches and diarrhea.

Field Notes:

Blue Flax

Blue Flax — **Flax Family**

Linum lewisii (*LEEN-um loo-IS-ee-eye*) — *Linaceae*

Linum is the Latin name for "linen," and *lewisii* refers to Meriwether Lewis of Lewis and Clark fame, American explorers who led an expedition to the Pacific Northwest from 1804-1806. Lewis served as the naturalist to the expedition.

Native to Canada, **blue flax** grows in dry, rocky areas north to Great Bear Lake and the arctic coast. The stem is 15-60 cm tall and arises from a woody base. The flowers grow close to the tips of the densely leafy stem. Each bloom lasts only a single day.

Leaves

Alternate; short; linear.

Flowers

Numerous; large; blue but occasionally white; five petalled, veined, centres yellow.

Plant Notes

Wild flax can often be found in disturbed soil and waste areas that are exposed to the open sun. This plant is closely related to cultivated flax, which has value because of its seeds. Flax oil, which is low in saturated fat, is an ingredient in many healthful foods, including breads and baked goods. Flax for fibre is typically grown in a climate of moist, cool summers and rich, porous soil. The seeds are sown close together to produce long, slender stems around 1.5 m high. Cultivated flax has been grown for centuries for its fibres for use in the manufacture of linen, thread, twine, writing paper and sailcloth.

Field Notes:

Bluebell

courtesy of Christian Bucher

Bluebell **Borage Family**
Mertensia paniculata (*mer-TENZ-ee-ah panee-KU-latah*) *Boraginaceae*
Named for German botanist, F.C. Mertens (1764-1831), the plant takes its species name from its compound inflorescence or panicle.

Bluebell is common in open woods, clearings and river banks throughout our area to just above the tree line. It grows along the Mackenzie River all the way to the arctic coast west of the Mackenzie River Delta. Bluebell has erect 20-70 cm stems.

courtesy of Christian Bucher

Leaves

Large, alternate; oval and long petioled near base of plant, becoming shorter farther up stem; surfaces rough and hairy.

Flowers

Drooping, bell shaped; pink in bud, turning blue when mature; sepals slightly hairy beneath.

Plant Notes

Leaves were used as a tobacco substitute by the South Slavey. Only the large leaves of non-flowering plants were gathered, sun dried and rubbed between the fingers to crush.

Field Notes:

Butterwort

Butterwort **Bladderwort Family**
Pinguicula vulgaris (*ping-GWI-kew-la vul-GAH-ris*) *Lentibulariaceae*
Pinguicula is from the Latin word for "fat," and refers to the greasy surface of the leaves. *Vulgaris* means "common."

Butterwort grows in damp, lime-rich soils by the edges of small brooks, ponds or seepages. It is a circumpolar plant of sub-arctic alpine regions. Delicate (4-12 cm tall) with a leafless stem, butterwort resembles purple violet.

Leaves

Basal; yellowish green; greasy to the touch.

Flowers

Long spurred; solitary; five petalled.

Plant Notes

The greasy leaves of this plant probably account for the common name of "butterwort." Wort is the old English name for "plant." Insects become trapped on the slimy leaves and become a food source. Like other carnivorous plants, butterwort has specialized features for trapping and digesting its food. Once trapped, insects struggle, which stimulates the plant to produce enzymes that break down the insect into a liquid that can be absorbed. The liquid nourishes the plant. One botanist has reported seeing as many as 500 blackflies stuck to the leaves of butterwort.

Field Notes:

Crowberry

Crowberry **Crowberry Family**
Empetrum nigrum (em-pee-trum NIG-rum) *Empetraceae*
Crowberry takes its name from the Greek words *em*, meaning "on," and *petros*, meaning "rock," referring to its habitat. *Nigrum* means "black," referring to the colour of the berries.

Crowberry is a low-growing, freely branching evergreen shrub that forms mats on top of rocks and acid soil. It is a wide ranging plant of the circumpolar world. The fruits are purplish black, shiny and smooth.

Leaves

Bristly, 5 mm long; look like evergreen needles, but are soft to touch.

Flowers

Inconspicuous; dark purple; growing in leaf axils.

Plant Notes

The South Slavey ate crowberries in early spring as well as after they ripened in August. When no water was at hand, the juicy berries were used to slake the thirst.

The Dogrib boil crowberry branches and roots to make a tea for mouth infections. The branches and berries are boiled for a tea to use for menstrual pain and the pain of childbirth, as well as for rashes. Crowberry is also boiled with tamarack and drunk to cure bad colds.

Field Notes:

Fringed Gentian

Fringed Gentian **Gentian Family**
Gentiana macounii (*gen-SHEE-ah-na ma-COON-ee-i*) *Gentianaceae*
Gentiana is from King Gentius, a 2nd century monarch of Illyria (modern-day Yugoslavia). *Macounii* is named for John Macoun (1831-1920), the pioneering Canadian botanist who set up the Dominion Herbarium.

The four-angled stems of **fringed gentian** grow up to 30 cm on gravelly beaches in the southern parts of NWT. Several stalks arise from each root with single or double flowers at tips. Fringed gentian may be easily confused with Raup's fringed gentian (*G. raupii*) because of the similarity in the flowers. Fringed gentian, however, has fewer leaves and its corolla has spreading lobes.

Leaves

Opposite; thin, entire; growing mostly toward base of plant.

Flowers

Four lobed; purple; conspicuously fringed at edges.

Plant Notes

A curious characteristic of gentians is the tiny fold in each petal that gives the flowers a sculptured look. This characteristic is obvious only upon close inspection.

Field Notes:

Harebell

Harebell — **Bluebell Family**

Campanula rotundifolia (*Cam-PAHN-ew-la row-tun-dee-FOAL-ee-ah*) — *Campanulaceae*

Campana is Latin for "bell;" *rotundifolia* means "round leaved."

Harebell grows from a freely branching rhizome in gravelly, peaty or rocky places. Because of the less tolerant growing conditions in NWT, we have only four members here of the otherwise large bluebell family. The erect or ascending 10-25 cm stems have the characteristic milky juice of the genus.

Leaves

Alternate, linear; clasping along stem; round at base of plant.

Flowers

Large, blue (sometimes white); 10-20 mm in diameter; 5 lobes fused to form bell.

Plant Notes

"Harebell," apparently, is a misspelling of "hairbell," which refers to the delicate, thin stems that hold up the flowers. As flowers open, they droop downward to protect pollen and nectar from rain.

Dene peoples of the boreal forest region use the roots of harebell in a tea for flu, fever and lung or heart trouble. It also has anti-inflammatory and antibacterial properties.

Field Notes:

Hedge-nettle

Hedge-nettle **Mint Family**
Stachys palustris (*STAY-kis paw-LUSS-tris*) *Labiatae*
Stachys means "spike;" *palustris* means "of wet areas."

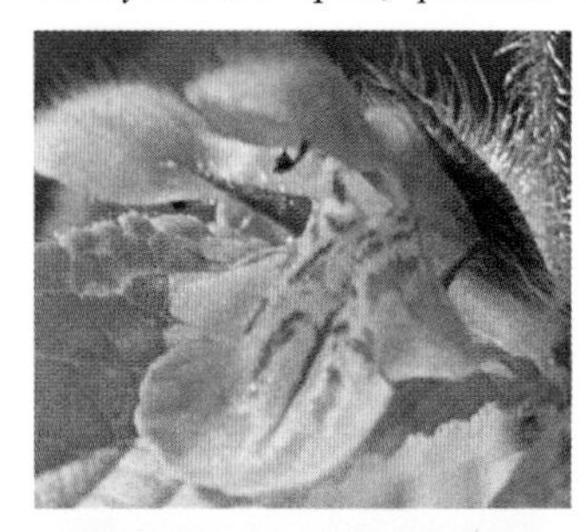

Rising 60 cm on a single straight stem from a slender rhizome, **hedge-nettle** bears a striking resemblance to wild mint, although on a larger scale. Unlike mint, hedge-nettle has no scent when its leaves are bruised. Both plants can be found in moist meadows and riverbanks in the circumpolar world.

Leaves

Opposite; lance shaped; toothed on margins; pointing upwards.

Flowers

Small; strongly two lipped, with lighter and darker spots of colour; in whorls in upper leaf axils continuing in a short, interrupted spike.

Plant Notes

The roots of hedge-nettle are edible and said to be quite tasty, eaten as is or cooked. In times of need, the roots were also ground and used to make flour.

Used for many years as a poultice and in ointments and syrups to stop bleeding, hedge-nettle has fallen into disuse because there is no research to support its medicinal benefit.

Field Notes:

Jacob's Ladder

Jacob's Ladder **Phlox Family**

Polemonium pulcherrimum (*po-lee-MO-nee-um pul-CHERR-i-mum*) *Polemoniaceae*

Polemonium is the Greek name of this plant. *Pulcherrimum* means "very handsome."

Jacob's ladder is a bright blue to violet roadside beauty that ranges widely from Alaska, Yukon, and northwestern NWT to the foothills of Alberta and BC. It lines the sides of the Dempster Highway from Inuvik to Fort Macpherson. Blue-flowered plants are much less common in the North than plants with white or yellow flowers so they tend to attract one's attention.

Leaves

Alternate; pinnate; up to 14 pairs of oval leaflets.

Flowers

Blue or purplish; five petalled; tube yellow; stigma divided into 3 curled sections at tip.

Plant Notes

Jacob's ladder is so-named because of the ladder-shaped arrangement of its leaves. Jacob's ladder is a biblical story about the ladder from earth to heaven that Jacob saw in a dream (Genesis 28:12). The terminology is also applied to a ladder of rope or wire used on a ship.

Field Notes:

Liquorice Root

Liquorice Root **Pea Family**
Hedysarum alpinum (*hay-DEE-say-rum al-PIE-num*) *Fabaceae*
Hedysarum is from the Greek and means "sweet smelling." *Alpinum* means "alpine."

With fewer and more erect stems than wild sweet pea (*H. mackenzii*, p. 85) **liquorice root** also grows taller, up to 60 cm from a large, fleshy root. This species is lime-loving as well, but it also grows in sand and gravel.

Leaves

Green; smooth; 9-13 pairs of lance-shaped leaflets.

Flowers

Pink or pale purple; tend to droop downward; scentless.

Plant Notes

The Dene of the South Slave would go to the mountains in September to dig this plant. They used sticks to dig down to the thick, horizontal roots. Soup was made from the fresh roots; stored roots were boiled with meat or sliced to be fried in grease. The taste is said to resemble young carrots.

Sore eyes were treated with this plant by the Dene of the South Slave. They burned small pieces of the sun-dried roots over a small fire and trapped the smoke with a blanket over the head. The plant contains compounds, among others, that are anti-inflammatory and antiviral.

Field Notes:

Philadelphia Fleabane

Philadelphia Fleabane **Sunflower Family**
Erigeron philadelphicus (*e-ri-GE-ron fill-a-DELL-fee-kus*) *Asteraceae*
Erigeron comes from the Greek *eri*, meaning "spring" or "early," and *geron*, meaning "old man," referring to the fluffy white seed heads. *Philadelphicus* means "of the Philadelphia region."

Philadelphia fleabane ia a wide-ranging boreal plant growing to a height of 20-100 cm on a leafy stem rising from a weak, fibrous root. The flowers grow in clusters at the stem tips. It can be found along moist river and lake shores, roadsides and clearings.

Leaves

Oblong; edges may be toothed or smooth.

Flowers

Pale purple; 10-15 mm diameter; ray florets numerous, narrow; disk yellow, dense.

Plant Notes

Philadelphia fleabane was used by Ojibwa peoples in a tea for fever. The pulverized flowers were also used to cause sneezing to loosen a head cold.

At one time, fleabanes were brought into houses to discourage fleas, which accounts for its common name. Although scientific research has shown that fleabanes can be used as an insecticide, modern sanitation makes this old folk use obsolete.

Fleabanes are often confused with asters. Fleabanes, however, bloom in spring and early summer; asters bloom in late summer or fall.

Field Notes:

Wild Mint

Wild Mint **Mint Family**
Mentha arvensis (*MEN-tha ar-VEN-sis*) *Labiatae*
Mentha is the Latin name for mint, derived from Greek mythology. *Arvensis* means "of the fields."

Wild mint grows in meadows, moist ditches, along river banks and lake shores. It is the only true native mint in North America; all other species have been introduced. Wild mint is a circumpolar plant, growing in the SW District of Mackenzie and north to Great Bear Lake. Wild mint is about 30 cm high with a square, purplish stem. When bruised, leaves give off a distinct minty aroma.

Plant Notes

Mint was used during Medieval times as a strewing herb, placed on floors in places of recreation, rest and repose. Mint leaves can be used as flavourings in drinks and also in cooking, especially Asian food.

The young greens are recommended as a potherb and can be chopped and mixed with dandelion greens. Mint is still used in a medicinal tea by the Dene of the boreal forest region. It is said to relieve a host of ailments, including bad breath, hiccups, coughs, headaches, fevers, upset stomach and high blood pressure. The leafy stem and flowers can also be used medicinally to pack into the nostrils as a treatment for severe nosebleed.

Leaves

Opposite; edges serrated; slightly hairy.

Flowers

4 mm long; clustered in whorls in leaf axils.

Field Notes:

Wild Sweet Pea

Wild Sweet Pea **Pea Family**

Hedysarum mackenzii (*hay-DEE-say-rum mac-KEN-zee-eye*) *Fabaceae*

Hedysarum is Greek for "sweet smelling;" *Mackenzii* is named after Alexander Mackenzie.

Wild sweet pea grows 15-35 cm high in lime-rich clays and gravels, especially along bodies of water. Many stems arise from the woody base of the plant, which is held firmly in the ground by a thick, fibrous tap-root.

Leaves

5-13 paired, linear leaflets,1.5 cm long; green and smooth above, with silvery hairs on underside.

Flowers

Inflorescence elongated; 5-25 pea-like flowers; sweetly scented.

Fruit

Joined seed pods (loments) characteristic of the genus.

Plant Notes

It is the scent and size of the flowers that help distinguish this nonedible species from the similar, yet edible, species called liquorice root (*H. alpinum*, p. 82). Wild sweet pea is considered poisonous, but liquorice root is edible. If you cannot tell these two species of *Hedysarum* apart, it is best not to experiment with either.

Field Notes:

Alpine Bearberry

Alpine Bearberry **Heath Family**
Arctostaphylos alpina (*ark-toe-STAAH-fil-os al-PIE-na*) *Ericaceae*
Arctostaphylos stems from *arctos*, which means "bear," and *staphyle*, which means "bunch of grapes." *Alpina* means "of alpine regions."

Alpine bearberry grows on rocky tundra in the arctic regions and in the boreal forest. This close relative of kinnikinnick has a similar, mat-forming habit, and also has edible, rather tasteless, fruit. With its highly noticeable scarlet leaves and luscious-looking black berries, it is definitely a herald of the coming winter. It accounts for the spectacular colour of the barrens in late August and September.

Leaves

Round-tipped; deeply toothed; wrinkled; not fleshy, but thin and somewhat leathery; turning scarlet in autumn.

Flowers

Small, urn-shaped; yellow-green; appearing early in season before leaves unfold.

Fruit

Shiny, black berries.

Plant Notes

The Tanaina of Alaska used the leaves of alpine bearberry in the early days to make a face powder. Leaves were picked in fall when they are bright red. After drying, the leaves were ground into powder, mixed with lard and applied to the face. Inuit of Baffin Island also made tea from the older leaves of the plant.

The closest relative of alpine bearberry is *A. rubra*, which has shiny red berries. The leaves of this species do not turn red, but fall off in autumn. It is more common in the southwestern area of NWT.

Field Notes:

Arrow-grass

Arrow-grass **Arrow-grass Family**
Triglochin maritima (*try-GLOW-chun mare-e-TY-mah*) *Juncaginaceae*
Triglochin is Greek for "three-pointed," referring to the fruit of other species of this genus. *Maritima* means "of the sea," referring to its habitat.

Tall, green and stiff, **arrow-grass** grows in salt flats, alkaline and boggy areas, and river meadows as far north as the tree line or slightly beyond. It grows to a height of 30-40 cm on a round, grooved, coarse stalk.

Leaves

Long, linear, somewhat fleshy; growing from base of plant, which is often under water.

Flowers

Very small, greenish; growing along length of stem, becoming more densely clustered toward tip; tightly attached to stem by short stalks, turning brown and dropping off readily at end of season.

Plant Notes

This plant is sometimes called seaside arrow grass, which is somewhat of a misnomer. The seaside here is only on the Beaufort, but I have found the plant even in gravelly areas that flood in spring.

Although considered poisonous because of its cyanide content, some of the Coastal tribes of British Columbia were known to eat the leaves and seeds of some species. Leaves were harvested in the spring and seeds were parched and ground into flour by several Aboriginal groups of western Canada. Roasted seeds were also used by pioneers as a coffee substitute.

Field Notes:

Bog-rush

Bog-rush **Rush Family**
Juncus arcticus (*JUNG-kus ark-TI-cus*) *Juncaceae*
Juncus is the Latin name for "rush;" *arcticus* means "of the arctic."

Bog-rush, also called wire rush, grows in gravel areas and sandy shores of lakes and rivers in the low-arctic region. About 25 cm high, bog-rush has smooth, stiff and wiry stems arising from the base all in a row, like the teeth of a comb. This horizontal rooting habit allows the plant to spread with ease. Like grasses, the stems of rushes are round. Unlike grasses, however, the stems are solid and unjointed.

Leaves

Yellowish-brown, sheathing stems from base of plant.

Flowers

Many, tiny, growing in clusters that appear to be on one side of stem only; covered by 6 brown-to-purple flower scales, giving the appearance of purplish flowers.

Fruit

At centre of flower; black, roundish capsule, which contains seeds.

Plant Notes

A similar-looking species, *J. balticus*, has been used in basketry by Native Americans. The blooming stems produce dyes ranging from ivory to ochre, green and yellow.

Field Notes:

Bur-reed

Bur-reed — **Bur-reed Family**

Sparganium. spp. (*spar-GAY-nee-um*) — *Sparganiaceae*

Sparganium is from the Greek word *sparganon*, which means "swaddling band."

Bur-reed is a tall (0.5-1.5 m) cat-tail like plant with long, graceful leaves that overtop the flowers. There are five species of bur-reed growing in Northwest Territories. They are very similar in appearance, except for differences in their tiny, one-seeded fruits. Bur-reed grows from rhizomes in shallow bog ponds, quiet lakes and river edges. Like grasses, bur-reeds have slender, leafy stems.

Leaves

Linear, sheathing stem; twisting somewhat toward top of plant.

Flowers

Male and female; female flowers form small, spiky ball; male flowers in much smaller heads at top of stem; heads appear to be growing out of female flowers, but are just farther up the stem.

Plant Notes

Bur-reed was just one of the thrilling finds on a trip to the Slave River Delta in July. As we motored up and down the channels of the delta, we marvelled at the lush vegetation, from the three metre tall horsetails (*Equisetum* spp. p. 6) to the towering sedges (*Carex rostrata*). At the very end of the first day on the channels, I saw a large plant that looked like a cat-tail without the "tail." As we got closer, I realized it was probably a bur-reed, which I knew only from an illustration in Porsild and Cody. I had longed to find this plant because of its unusual flowering heads shaped like burrs. The prickly seed heads look like tiny rambutan, a prickly coated fruit from Indonesia.

Field Notes:

Cat-tail

Cat-tail **Cat-tail Family**
Typha latifolia (*TY-fa lat-e-FOAL-ee-ah*) *Typhaceae*
Typha is from the Greek *typhos*, which means "marsh;" *latifolia* means "broad-leaved."

Cat-tail grows to 1-2 m at the edge of ponds, streams, sloughs and marshy areas. Cat-tail has a distinctive fuzzy "tail" at the tip of the dense, brown spike.

Leaves

Flat, linear, 1-2.5 cm wide; may be longer than stem.

Flowers

Numerous, tiny; male flowers at tip of spike, turning gold with pollen when mature; numerous female flowers in lower dense brown spike, turning velvety brown when mature.

Plant Notes

Cat-tail is a versatile food plant. The tender inner core of the spring stems can be eaten raw or cooked. The green immature flower spikes can be substituted for corn on the cob; pollen from the male flowers can be collected and used in pancakes, muffins and biscuits; and the starchy rootstock can be pounded and used for flour. The South Slavey dug the rhizomes in the fall to be eaten raw or fried in grease. The root stalks of cat-tail are also eaten by muskrats.

Cat-tail also has medicinal uses. Mature female flowers can be mashed and used in a salve for cuts and burns. The Dogrib burn the cat-tail stalk and use the ashes for skin rash. The South Slavey gathered the fluff from mature cat-tails and mixed it with moss in baby bags to increase absorbency and to keep babies warm.

Field Notes:

Common Great Bulrush

Common Great Bulrush — **Sedge Family**

Scirpus validus (*SKIR-pus vah-LEE-dus*) — *Cyperaceae*

Scirpus is Latin for "bulrush;" *validus* means "strong."

Great bulrush can be seen growing up to 2 m at the perimeters of numerous ponds and wet areas as far north as Norman Wells and along the highway from Fort Providence to Yellowknife. This is the tallest bulrush growing in our area, which makes it easy to identify.

Leaves

Represented by sheaths, each with a stiff pointed tip.

Flowers

Drooping, 5-10 cm clusters from short stalks.

Plant Notes

Like cat-tails (*Typha latifolia*, p. 90), bulrushes can provide food all year: the new shoots are eaten raw or boiled, the pollen can be collected in bags in spring, and the seeds can be collected in fall. The young roots of the bulrush are eaten as a vegetable or sugar source while old roots can be pounded for flour.

The stem pith also finds use as a wound dressing to stop bleeding. The strong stems of this species have been used to make many household items such as chair seats, mats, mattresses, baskets and thatching for roofs.

Field Notes:

Duckweed

Duckweed **Duckweed Family**
Lemna minor (*LEM-nah MINE-er*) *Lemnaceae*
Lemna is the Latin word for plants with one root. *Minor* means "smaller." Duckweed is eaten by ducks, thus its common name.

Duckweed has a single root but is stemless. It grows on the surface of stagnant pools and ponds throughout NWT. It can cover large areas and looks like green scum. If you scoop it up, you will see that it is not at all slimy, just smooth and wet.

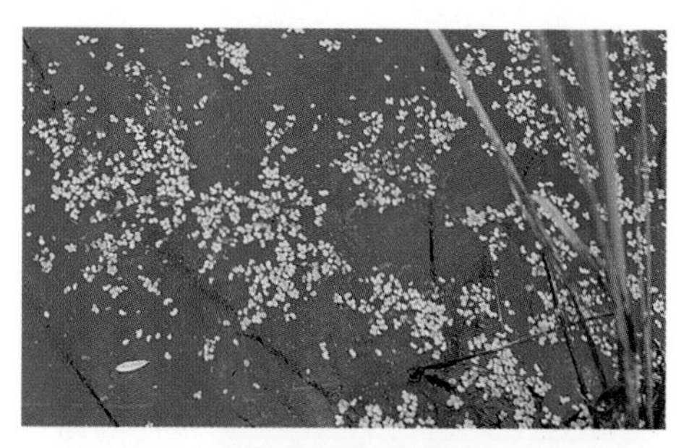

Leaves

No leaves but tiny (2-5 mm), green, disk-like floating bodies called thalli, each with a single root suspended in water.

Flowers

Rare; minute, easily overlooked because of size.

Plant Notes

Duckweed plants multiply mainly from tiny buds that form along the edge or upper surface of the parent thallus.

The South Slavey used duckweed as an indicator plant. They knew if they drank the water in which duckweed grew, an itchy rash would develop on the hands and feet.

Field Notes:

Dwarf Birch

Dwarf Birch **Birch Family**
Betula pumila (*be-CHEW-la pu-MEE-la*) *Betulaceae*
Betula is the Latin name for birch; *pumila* means "dwarf."

Dwarf birch grows to 2 m mainly in woodland bogs and moist areas of the boreal forest. Shrubby in habit, dwarf birch has many stems. The branches are often dotted with raised pores called lenticels, which gives them a nubby texture. The bark is reddish brown or dark grey. Although female and male catkins grow on the same plant, they mature at different times and are thus pollinated by other plants.

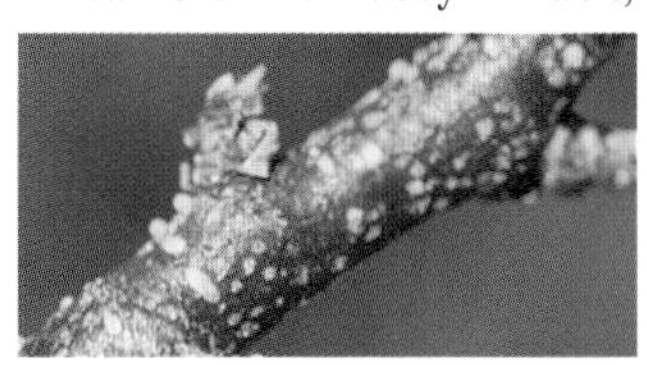

Leaves

Alternate, 1-4 cm; rounded at base; serrated on edges; green on both sides, but slightly paler underneath.

Flowers

Female catkins round and woody, maturing first, before leaves appear; male catkins slender, drooping; releasing pollen into wind.

Plant Notes

Cones of dwarf birch were used by native North Americans for chronic nasal infections. Cones were roasted on the coals of a low fire and the fumes were inhaled. Dene peoples of Saskatchewan chew the fresh twigs and put them on a deep cut to stop bleeding.

Ground birch (*B. glandulosa*) is of similar habit, but shorter. The two species are easily confused and at times hybridize, compounding the confusion.

Field Notes:

Fox-tail Barley

Fox-tail Barley **Grass Family**
Hordeum jubatum (*hor-DEE-um you-BAY-tum*) *Poaceae*
Hordeum is the Latin name for "barley;" *jubatum* means "crested, with a mane."

Fox-tail barley grows as a weed in waste places north to the Mackenzie Delta. Clumps of several, erect stems, 30-60 cm high, can be seen along roadsides and in gravelly places. It is the graceful arch of the spikelets that gives the plant its species name. The spikelets catch each breath of wind as the plants shimmer from the roadside.

Leaves

Grey-green; rough; flat.

Flowers

Spikes bristly, nodding; awns pale green when immature, turning purplish when mature and yellow when seeds are ripe.

Plant Notes

When fox-tail barley matures, the awns break into units of three, each with one-flowered spikelets. The awns are light and easily carried by the wind. The sharp, barbed·tip readily attaches the awn to the fur of passing animals or the clothing of passing people. In this way, the seeds can hitch-hike great distances; this accounts for its wide spread.

The Saskatchewan Dene harvest foxtail barley as food in early summer when the shoots are young and tender.

Field Notes:

Green Alder

Green Alder **Birch Family**
Alnus crispa (*AL-nus KRIS-pa*) *Betulaceae*
Alnus is the Latin name for "alder;" *crispa* means "crisped" or "curled," referring to the leaf surface.

Green alder is the common alder of wooded areas. It is shorter (3 m) than grey alder which grows up to 5 m. The young leaves and twigs are sticky to the touch. Flowers develop as male and female catkins on new twigs at the same time as the leaves.

Leaves

Oval; shiny; irregularly toothed.

Flowers

Greenish brown, forming catkins.

Plant Notes

Species of alder were used to make pipes by the Dene of the South Slave. Barking tools were made from alder, and were designed so that the working edge of the tool would curve around a tree trunk. Alder wood was also used for making buttons and lids for small birch bark containers. Small bows for birds and squirrels were made from the tough, flexible branches. Rotten alder wood served as an insect repellent: smouldering chunks were placed in the four corners of the sleeping area to keep insects at bay. A smouldering stick was carried and frequently waved about to keep bugs off. The Sahtu Dene use alder for sexual diseases, as well as rashes and itching.

Field Notes:

Northern Green Orchid

Northern Green Orchid **Orchid Family**
Habenaria hyperborea (*ha-bay-NAA-ree-a hy-per-BOR-ee-a*) *Orchidaceae*
Habenaria comes from the Latin *habena*, which means "thong" or "strap," referring to the shape of the lip; *hyperborea* means "of the far north."

Northern green orchid grows to 55 cm in wet and boggy places. The leafy stem is topped by scentless flowers. One word describes this plant: green. It tends to blend in with its surroundings, so it may prove elusive to the wild plant seeker. Its flowers resemble those of tway blade (*Liparis loeselii*), but the leaves do not. Tway blade has only two leaves.

Leaves

Alternate; lance-to-oblong shaped; sheathing stem, becoming bracts toward top of plant.

Flowers

Small, greenish, growing on both sides of stem in 2-10 cm spike.

Field Notes:

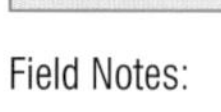

Mare's-tail

Mare's-tail — **Water-milfoil Family**

Hippuris vulgaris (*Hi-PEWR-ris vul-GAH-ris*) — *Haloragaceae*

Hippuris comes from the Greek *hippos* for "horse," and *oura* for "tail." *Vulgaris* means "common."

Mare's-tail is an aquatic, wide-ranging, circumpolar plant of shallow ponds and lakes. It grows from a fleshy, unbranched stem that rises from a creeping rhizome anchored in mud. The stem length varies from 10-50 cm, depending on depth of water. In deeper water, stems tend to become elongated. Mare's-tail is an attractive plant because of its symmetry and the way the whorls stand out from the stem. Mare's-tail can be confused with horsetail (*Equisetum* spp., p. 6) because of their similar habitat and somewhat similar appearance.

Leaves

In whorls of 6-12; submersed leaves longer than emergent leaves.

Flowers

Minute; inconspicuous.

Plant Notes

I found mare's-tail in Yellowknife in the ponds behind the Northern Frontier Visitors' Centre and at Niven Lake. In addition to its symmetry and the nature of its whorls, the plant looks particularly attractive in fall when it seems to twist up and out of water as the water recedes before winter freeze-up.

Field Notes:

Marsh Reed Grass

Marsh Reed Grass **Grass Family**
Calamagrostis canadensis (*kal-a-ma-GROS-tis KAN-a-den-sis*) *Poaceae*
Calamagrostis is derived from the Greek *calamus* meaning "reed," and *agrostis* meaning "grass." *Canadensis* means "of Canada."

Marsh reed grass, also called blue joint because of its blue nodes, is a wide-ranging species of the boreal forest, common along the Mackenzie River to its delta. Because of its height (50-150 cm) and large, purple or green plumes, it is one of the easiest grasses to identify. Marsh reed grass grows in tufts from a creeping rhizome.

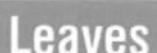

Leaves

Soft, flat, turning brittle and bending back as the plant dries out at end of season.

Flowers

Panicles feathery, drooping; spikelets one-flowered.

Plant Notes

Marsh reed grass is a food for moose during the month of September. The South Slavey used grasses from the *Calamagrostis* genus as drinking straws because the stalks are stiff and hollow. Other Aboriginal groups have used marsh reed grass to stuff mattresses and to line food storage pits.

Field Notes:

Northern Comandra

Northern Comandra **Sandalwood Family**
Geocaulon lividum (*gee-o-CALL-on li-VEE-dum*) *Santalaceae*
"*Comandra*" is said to be derived from the Greek *kome* for "hair," and *andros* for "man," thought to refer to the hairy bases of the stamens. The greenish-purple calyx is thought to be the source of the species name, *lividum*, which means "bluish" in Latin.

Commonly seen in mossy areas and open, moist woods, **northern comandra** is one of only two species of the sandalwood family growing in the North. Both species grow from creeping rhizomes, which are parasitic on the roots of other plants, including bearberries. Stems are erect and 10-25 cm high. The plant is more noticeable in the early autumn when the leaves turn from green to deep, rusty red.

Leaves
Alternate; lance shaped; numerous.

Flowers
Greenish; tiny; axillary.

Fruit
Bright orange berries.

Plant Notes
The Dogrib boil the berries and use the tea to cure respiratory problems. The Outer Inlet people of Alaska use the chewed, raw berries, or an infusion of the roots, for stomach trouble, sore throats and other ailments. They also use the fresh leaves, which are partially chewed or mashed, to apply to sores as a poultice.

Field Notes:

One-sided Wintergreen

One-sided Wintergreen **Wintergreen Family**
Pyrola secunda (*PI-ro-la see-KUN-dah*) *Pyrolaceae*
Pyrola comes from *pyrus* (pear) because the two plants have similar leaves; *secunda* means "side-flowering."

One-sided wintergreen is a delicate woodland species that grows north beyond the limit of trees. Growing up to 20 cm from a long rhizome, the plant has tiny bracts on the stem and several basal leaves typical of the *Pyrola* genus. The flowers are shaped like tiny green bells and the projecting style can almost be visualized as the clapper of a bell.

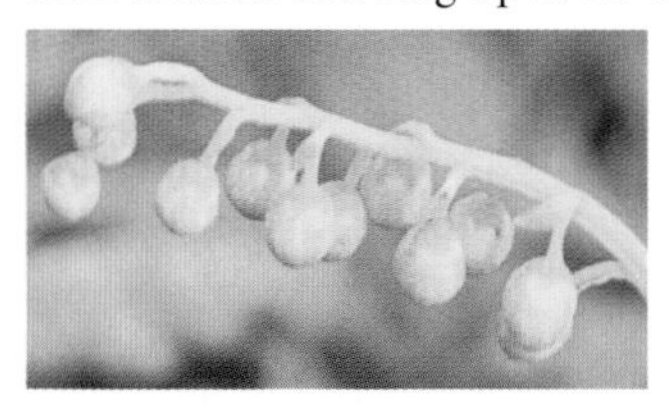

Leaves

Oval, slightly pointed, leathery to the touch; leaf margins toothed, but a hand lens may be needed to see this feature.

Flowers

Several, small, greenish; growing on upper part of stem in one-sided row.

Plant Notes

Pyrolas have properties that have been reported to reduce pain and inflammation and enhance circulation. The Dene of Saskatchewan mashed the leaves with lard to put on a cut to staunch the flow of blood. Leaves were also reported to be good for toothache. A leaf infusion can be used to wash sore eyes.

I found this plant in the woods at Jack Fish Lake Park in Norman Wells. This little wintergreen, growing under a bush and as green as the surrounding forest plants, was difficult to spot.

Field Notes:

Paper Birch

Paper Birch **Birch Family**

Betula papyrifera (be-CHEW-la pa-pi-RI-fe-ra) *Betulaceae*

Betula is the Latin name for "birch;" *papyrifera* means "paper-bearing," referring to the paper-like bark.

Paper birch is the common white-barked birch of the western boreal forest. Reaching heights of 30 m, it can be found on Precambrian rocks and in acid and peaty soils. The bark has obvious lenticels that allow the exchange of gases between the tree and the environment. The bark easily pulls off in horizontal strips.

Leaves

Lance shaped, irregularly toothed; bright green, turning brilliant yellow in fall.

Flowers

Greenish-brown male and female catkins; produced before leaves appear.

Plant Notes

Paper birch has traditionally been a valuable food plant for Aboriginal peoples of this area. The inner bark can be scraped off into noodle-like strips and used as a starvation food or boiled to make a beverage. The Dene of the South Slave collected the sap and drank it as a health food or boiled it down for sap. The Sahtu Dene use the outer bark for kidney, bladder and stomach problems. The inner bark is also used for kidney disease as well as skin rashes, cancer and liver disease.

Field Notes:

Plantain

Plantain — **Plantain Family**

Plantago major (*plan-TAY-go may-jur*) — *Plantaginaceae*

Planta means "sole of the foot;" *major* means "larger."

Plantain grows as a weed near many settled areas in the North. It is common along roadsides and travelled areas. Its seeds have been known to lay dormant for decades under sidewalks, then sprout between the cracks.

Leaves

Oval, abruptly tapering; ribbed with 5-7 obvious veins; edges toothed.

Flowers

In a dense, narrow spike; flowers tiny.

Plant Notes

Plantain has strong healing power. The fresh leaves can be pounded into a paste and used to stop bleeding. Plantain leaves can also be used as a poultice for insect bites, boils, carbuncles and tumours.

Aboriginal peoples named common plantain "white man's foot" because it grew only where white men had travelled.

Field Notes:

Richardson's Pond Weed

Richardson's Pond Weed **Pondweed Family**

Potamogeton richardsonii (*Pot-a-mo-JEE-tun Richard-SOWN-ee-i*) *Potamogetonaceae*

Potamogeton is from the Greek words for "river" and "neighbour." *Richardsonii* refers to John Richardson, botanist to the Franklin Expedition of 1819.

Richardson's pond weed is widespread across our region in ponds, lakes and slow-moving streams up to 3 m deep. It can be seen floating just below the surface where the water is clear. Most of the plant is submerged, from its slender rhizomes up to the leafy round stem. The dense flower stalk pokes its head straight out of the water. The whole plant is brownish-green and has a slimy texture when wet. When dried and pressed, it looks more green than brown, and is papery to touch.

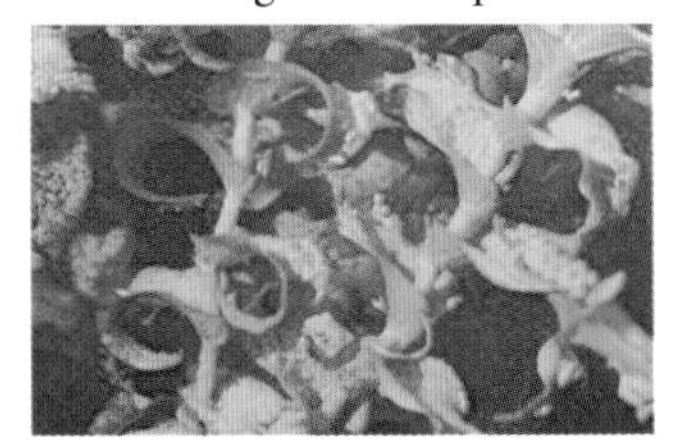

Leaves

Numerous; lance shaped, 3-12 cm long; translucent with prominent vertical veins and wavy edges.

Flowers

Tiny; broccoli shaped; growing around top of stalk.

Plant Notes

There are many species of pond weed in NWT, some of which serve as food for muskrats and waterfowl. When seen floating just below the surface, it looks like kelp from the sea with its wavy, translucent leaves that move gently under water with any surface disturbance. This plant was gathered for me by my husband and daughter at Vee Lake just outside Yellowknife.

Field Notes:

Soapberry

Soapberry **Oleaster Family**
Shepherdia canadensis (*she-PERD-ee-ah kan-a-DEN-sis*) *Elaeagnaceae*
Shepherdia is from the name of British botanist, John Shepherd (1764-1836); *canadensis* means "Canadian."

Soapberry grows to a height of well over 1 m in open woods and lime-rich soils to just beyond the tree line. It is wide ranging throughout northern North America. Branches grow either straight up or curve upward from the base. The leaves and twigs have a scurfy coating that is rough to the touch. Plants may be male or female.

Leaves

Opposite; long; oval; undersides have silvery hairs.

Flowers

Small; brown; growing in leaf axils.

Fruit

Juicy, red berries.

Plant Notes

As food, soapberries can be whipped with water and sugar to make a fluffy topping for desserts. They are high in vitamins A, B and E. Eaten on their own, the berries are astringent and bitter, although very juicy. Cooked with grease or sugar, or pressed into cakes and smoked, soapberries have served as food to many different Aboriginal groups. Soapberries have been used in many traditional medicines. The Sahtu Dene, for example, used it to relieve constipation. It can be used as a purgative as well as treatment for tuberculosis. Soapberry tea, used topically, is good for cuts swellings, impetigo and as a treatment for arthritic aches and pains.

Field Notes:

Water Sedge

Water Sedge — **Sedge Family**

Carex aquatilis (*KARE-ex ah-QWA-ty-lis*) — *Cyperaceae*

Carex is the Latin name for "sedge;" *aquatilis* means "growing in water."

This grass-like member of the sedge family grows in dense tufts along the edges of sloughs, rivers or ponds. In some areas, it forms extensive stands. **Water sedge** has slender, yet solid, triangular stems. It grows up to 1 m in height. The horizontal stolons are scaly, stout and yellowish in colour.

Leaves

Many; flat or channelled; 2-5 mm broad; sometimes waxy; may twist and turn downward.

Flowers

Male spikes 1-2, blackish-coloured, terminal; female spikes 2-5, larger, cylindrical, lighter-coloured, cupped in long leaf-like bracts, which grow taller than either male or female spikes.

Plant Notes

The roots of water sedge have been used by the Dene peoples of the boreal forest region as part of a compound tea for intestinal problems, while the Tanaina of Alaska used members of this plant family to make brooms. They also wove sedges into mats and baskets. The South Slavey used water sedge as a dye for porcupine quills.

Field Notes:

Willow

Willow | **Willow Family**
Salix spp. (*SAY-liks*) | *Salicaceae*
Salix is Latin for "willow."

Willows grow just about everywhere in the North: close to streams, on riverbanks, in open forest, moist meadows, river flats and stony tundra. Willows are usually shrubby, with straight and smooth stems, but some can grow into trees up to 3 m. Because willows sprout easily and hybridize readily, it can be difficult to identify the species.

Leaves

Alternate, in clusters on stem, narrow, pointed.

Flowers

Male and female catkins growing on separate plants.

Plant Notes

Infusions made from willow bark have been used to treat pain since the time of the Ancient Greeks. Willow contains salicin, from which our present-day pain relievers were derived.

Willow has been used by local Aboriginal peoples as a pain reliever as well. It has also been used as food and for making rope and tools.

Field Notes:

Glossary

Although I have strived to keep this book as free of jargon as possible, in some cases it was necessary to use the specialized terminology that describes plants and their properties. The following list explains terms that may not be part of the everyday language of readers.

Aerial parts The parts of a plant that grow above ground or water surface.

Alkaloid Part of the basic chemical structure of some plants that are usually used for drugs.

Anthers Sac-like structures on stalks of stamens. Anthers contain pollen grains.

Awn A slender bristle.

Axil The upper angle between a leaf (or scale) and the stem on which it is borne.

Axillary Growing in axils of leaves.

Basal rosette Arrangement of leaves that appears to be growing directly from the ground, but in fact, is attached to an underground stem or rhizome. Rosette refers to the rose-shaped arrangement of leaves.

Bilateral Refers to a flower that is divisible into equal halves along only one plane; same as "irregular," e.g., round-leaved orchid, p. 31.

Bract A leaf or a structure in the axil of which a flower is formed.

Calyx A collective term for the sepals of a flower.

Cambium A narrow band of cells in the inner bark of woody stems. It produces bark on the outside and wood on the inside.

Cauline Pertaining to leaves that are attached to the above-ground part of stem.

Leaf Arrangements

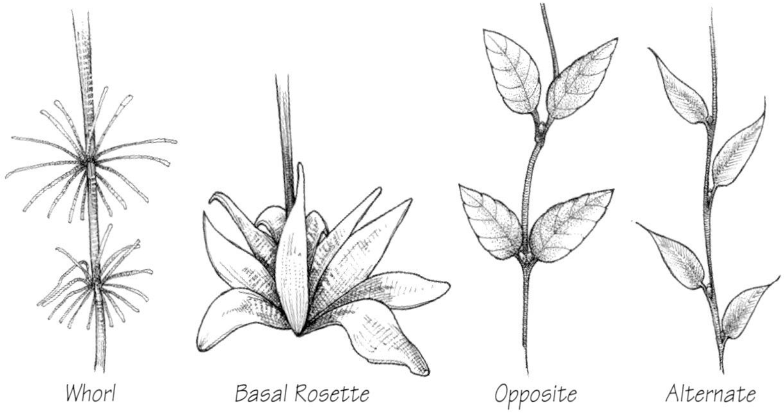

Whorl *Basal Rosette* *Opposite* *Alternate*

Corolla A collective term for the petals of a flower.

Decoction A process by which substances are extracted by boiling. Usually considered a stronger preparation than an infusion.

Disk florets Small tubular flowers in the central part of the flower head in sunflower family.

Diuretic A medicine that promotes the excretion of urine.

Emergent Plant parts that rise above the water surface

Family A group of genera having some common characteristic.

Frond The leaf of a fern or palm.

Genus A group of species having some common characteristics; ranking between family and species. (Plural: genera)

Head A dense flower cluster with ray and/or disk florets, as in sunflower family, eg., Philadelphia fleabane, p. 83.

Helmet The hood-shaped fused upper petals in some flowers.

Hybrid The offspring of two plants that are genetically different from each other.

Hypanthium A structure formed by the union of bases of sepals, petals and stamens. It may be shaped like a disk, cup or tube; e.g., rose hip is a cup-shaped hypanthium.

Imperfect Refers to a flower that has either stamens or pistils. Also called unisexual.

Inflorescence Any flower cluster found on a plant. When there is only one flower on a plant, the flower is called solitary, eg., one-flowered wintergreen, p. 27.

Infusion Medicinal liquid prepared by soaking plant parts in water.

Leaf Shapes

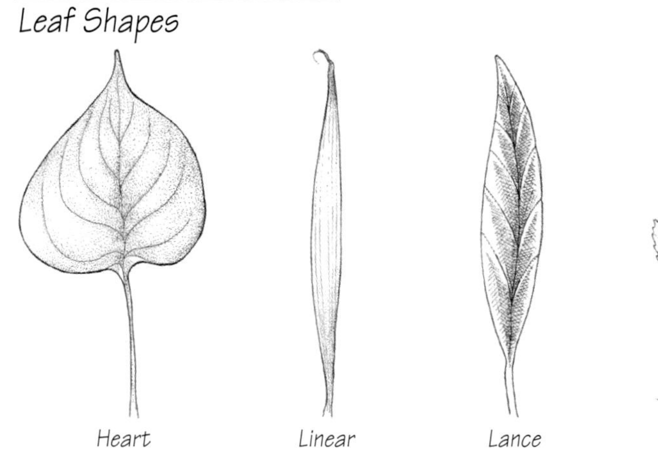

Heart *Linear* *Lance* *Frond*

Inferior Refers to an ovary that appears to be below sepals; here sepals, petals and stamens arise from its top.

Involucre One or more whorls of bracts that subtend an inflorescence.

Irregular Flower type in which the flower can only be divided into two symmetrical halves. Also called bilateral. (See also "regular.")

Lenticel A pore (often appearing as a crack) in the bark of woody plants that allows exchange of gases between a stem and the atmosphere. The bark of birch trees has large, elongated black lenticels.

Linnaeus The Swedish botanist (1707-1778) who devised a system of naming plants that is still in use today.

Mordant A corrosive acid that fixes colours in dyeing.

Mycorrhiza A fungus-root association.

Nectary A glandular structure that secretes a sweet fluid (nectar).

Node The point at which a leaf is attached to a stem.

Panicle A compound inflorescence shaped like a pyramid in which the youngest flowers are at the tips of its branches, eg., bluebell, p. 75.

Perfect Refers to a flower that has both stamens and pistils. (See also "imperfect.")

Perianth Collective term for the sepals and petals of a flower; commonly used when these organs look alike, e.g., lilies or sedges.

Petiole Stalk of a leaf.

Pinna The individual part of a pinnately compound leaf or frond. (Plural: pinnae)

Leaf Shapes

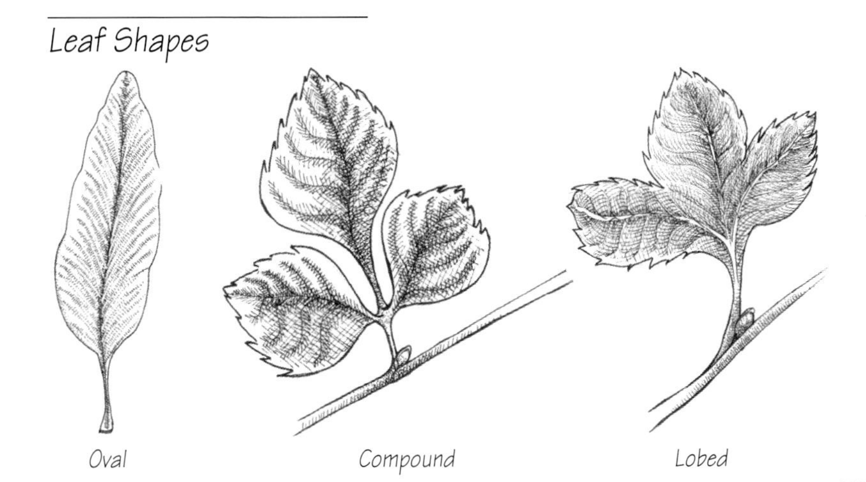

Oval *Compound* *Lobed*

Pinnate	Arrangement of leaflets on each side of a leaf axis, like a feather.
Pistil	The female reproductive organ of a plant. It consists of an ovary, a style and a stigma.
Pollen grains	Minute spores produced in anthers of stamens. Each mature spore produces two sperms.
Pollination	The process of transferring pollen from a stamen to a stigma.
Raceme	An elongated inflorescence in which the lowest flowers open first, eg., fireweed, p. 65.
Racemose	Like a raceme.
Ray florets	Strap-like flowers in the flower head of sunflower family.
Regular	Flower that can be divided in several symmetric halves. Also called "radial." (See also "irregular.")
Rhizome	Underground horizontal thick stem.
Salicin	A bitter white crystalline substance obtained from the bark of willows and poplars.
Scape	A leafless flower stalk.
Sepal	A unit of calyx that is the outermost part of a flower; it is usually green.
Simple stem	A stem that does not branch.
Spadix	A flower spike on a fleshy axis, eg., water-arum, p. 35.
Spathe	A large, often showy, bract that encloses or subtends an inflorescence, which is usually a spadix.
Spike	A raceme with sessile (stalkless) flowers, eg., northern green orchid, p. 96.
Spikelet	A small or secondary spike.

Root Structures

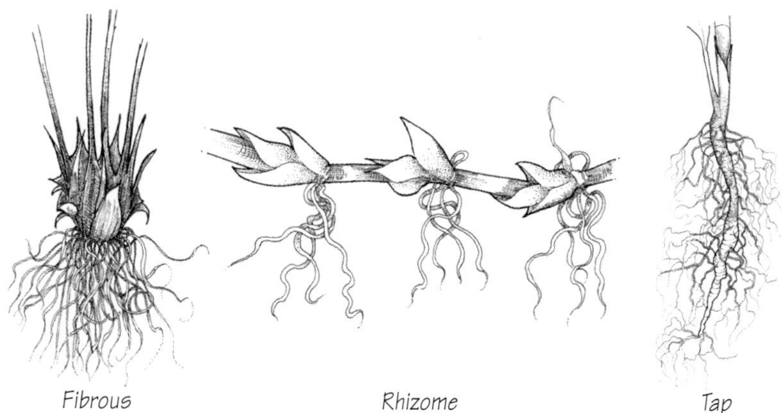

Spore A reproductive cell lacking an embryo. Spores are found in all plants. Fern spores are in small sacs on undersides of leaves. Spores of flowering plants are in stamens (pollen) and in immature seeds.

Stamen The male reproductive organ of a flower. Stamens are tipped by anthers, which contain pollen. The narrow stalk of a stamen is called the filament.

Stigma Receptacle for pollen at the tip of the style.

Staminodium A sterile stamen. (Plural: staminodia)

Stipules A pair of appendages attached at the base of a leaf.

Stolon A horizontal, often thin, stem, growing above the surface of the ground, and often rooting at the nodes.

Strobilus A cone-like aggregation of spore-bearing leaves or scales, as in club mosses or horsetails. (Plural: strobili)

Style Part of the female reproductive organ (or pistil) that links the ovary to the stigma.

Subtend To occur immediately below, as a bract subtending a flower; or an involucre subtending an inflorescence.

Tannin An astringent substance obtained from plant parts. Can be used to tan hides.

Tonic An invigorating medicine, meant to tone the system.

Umbel An umbrella-shaped flower cluster in which the stalked flowers arise from a common point, eg., water parsnip, p. 33.

Vascular plant A plant with specialized tissues for transportation of water and nutrients. All plants described in this book are vascular plants.

Whorl Three or more leaves arising from a single node.

Root Structures

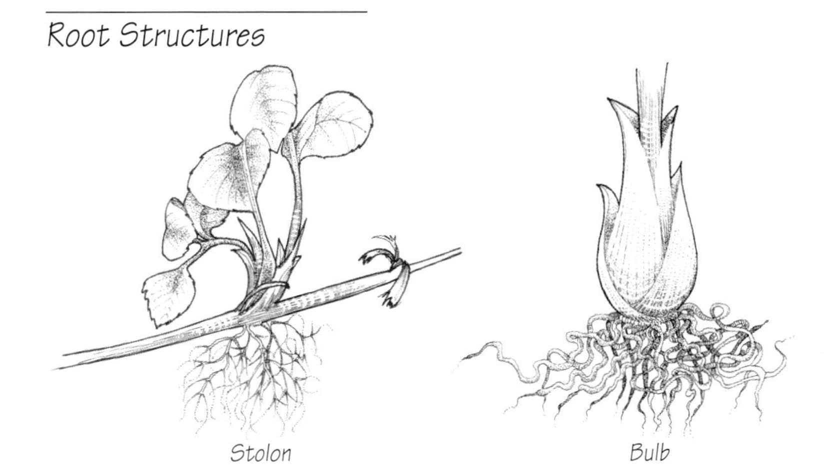

People from a planet without flowers
would think we must be
mad with joy the whole time
to have such things about us.

Iris Murdoch

Flowers are words
which even a baby may understand.

Arthur C. Coxe

Bibliography

Addison, Josephine. 1985. *The Illustrated Plant Lore: A Unique Pot-Pourri of History, Folklore and Practical Advice*. Sidgwick and Jackson, London.

Andre, Alestine and Fehr, Alan. 2001. *Gwich'in Ethnobotany: Plants Used by the Gwich'in for Food, Medicine, Shelter and Tools*. Gwich'in Social and Cultural Institute and Aurora Research Institute, Inuvik, Northwest Territories.

Angier, Bradford. 1978 *Field Guide to Medicinal Wild Plants*. Stackpole Books, Harrisburg, Pennsylvania.

Bailey, L.H. 1975 *How Plants Get Their Names*. Gale Research Company, Detroit, Michigan.

Bliss, Anne. 1976. *North American Dye Plants*. Charles Scribner's and Sons, New York, New York.

Castleman, Michael. 1991. *The Healing Herbs: The Ultimate guide to the Curative Powers of Nature's Medicines*. Rodale Press.

Chevallier, Andrew. 1996. *The Encyclopedia of Medicinal Plants*. 1996. Dorling Kindersley Ltd., London.

Clark, George H. and James Fletcher. 2000. *Farm Weeds of Canada, 2nd ed.* Algrove Publishing Limited, Ottawa, ON. (Reprint of original 1906 edition).

Coombes, Allen J. 1985. *Dictionary of Plant Names*. Timber Press, Beaver, Oregon.

Elias, Thomas S. and Peter A. Dykeman. 1990. *Edible Wild Plants: A North American Field Guide*. Sterling Publishing Co. Inc., New York, New York.

Farrow, Judy. 1993. *Arctic Plants of Baffin Island: Inuktitut Names and Traditional Uses* (Draft Document). Yellowknife, NT.

Fischer-Rizzi, Susanne. 1996. *Medicine of the Earth*. Rudra Press, Portland, Oregon.

Fitzgibbon, Agnes. 1972. *Canadian Wildflowers*. Botanical descriptions by Catherine Parr Traill. Facsimile edition, Coles Publishing Co., Toronto, Ontario.

Fitzharris, Tim. 1986. *Wildflowers of Canada*. Oxford University Press, Toronto, Ontario

Heatherley, Ana Nez. 1998. *Healing Plants: A Medicinal Guide to Native North American Plants*. Harper Collins Publishers Ltd., Toronto, Ontario.

Hemphill, John and Rosemary. 1984. *Herbs: Their Cultivation and Usage*. Blandford Press, London.

Hutchens, Alma.1991. *Indian Herbalogy of North America: The Definitive Guide to Native Medicinal Plants and Their Uses*. Shambala Publications Inc., Boston, Massachusettts.

Ingram, John. 1869. *Flora Symbolica*. F. W. Warne and Co., London.

Johnson, Derek *et al*. 1995. *Plants of the Western Boreal Forest and Aspen Parkland*. Lone Pine Publishing, Edmonton, AB.

Johnson, Karen L. 1987. *Wildflowers of Churchill and the Hudson Bay Region*. Manitoba Museum of Man and Nature, Winnipeg, MN.

Kari, Priscilla Russell. 1991. *Tanaina Plantlore, 3rd ed.* Alaska Native Language Center, Alaska Natural History Association and National Park Service.

Kerik, Joan. 1975. *Living off the Land: Use of Plants by the Native People of Alberta*. Alberta Culture Circulating Exhibits Program, National Museums of Canada Fund, Provincial Museum of Alberta.

Lamont, S.M. 1977. *The Fisherman Lake Slave and Their Environment: A Story of Floral and Faunal Resources*. M.Sc. thesis, Department of Plant Ecology, University of Saskatchewan, Saskatoon.

Lathrop-Smit, Hermine. 1978. *Natural Dyes*. James Lorimer and Company, Toronto, Ontario.

Liniger, Schuyler W. Jr. *et al*, eds. 1999. *The Natural Pharmacy. 2nd ed.* Prima Publishing, Rocklin, California.

Marles, Robin J. *et al*. 2000. *Aboriginal Plant Use in Canada's Northwest Boreal Forest*. UBC Press, Vancouver, BC.

McGrath, Judy. 1977. *Dyes from Lichens and Plants*. Toronto, Van Nostrand Reinhold.

Michael, Pamela. 1980. *All Good Things Around Us: A Cookbook and Guide to Wild Plants and Berries*. Holt, Rinehart and Winston, New York, New York.

Ody, Penelope. 1993. *The Complete Medicinal Herbal*. Dorling Kindersley, Ltd., London.

People of 'Ksan. 1980. *Gathering What the Great Nature Provided: Food Traditions of the Gitksan*. Vancouver, Douglas and McIntyre.

Phillips, Roger and Nicky Foy. 1990. *The Random House Book of Herbs*. Random House, New York.

Porsild, A.E. and W.J. Cody. 1980. *Vascular Plants of Continental Northwest Territories, Canada*. National Museums of Canada, Ottawa, Canada.

Potterton, David, ed. 1983. *Culpeper's Color Herbal*. Sterling Publishing Co., New York, New York.

Pringle, Laurence. 1978. *Wild Foods*. Four Winds Press, New York, New York.

Rohde, Eleanour Sinclair. 1969. *A Garden of Herbs*. Dover Publications Inc.

Royer, France and Richard Dickinson. 1999. *Weeds of Canada and the Northern United States*. The University of Alberta Press and Lone Pine Publishing, Edmonton, AB.

Schofield, Janice J. 1989. *Discovering Wild Plants: Alaska, Western Canada, the Northwest*. Alaska Northwest Books, Portland, Oregon.

Simmons, Ellen. 1999. *Report of Traditional Knowledge and Medicinal Uses of Plants from the Sahtu*. Resources, Wildlife and Economic Development, Government of the NWT, Norman Wells, NT.

Stearn, William T. 1983. *Botanical Latin*. Fitzhenry and Whiteside, Markham, Ontario.

Turner, Nancy J. 1995. *Food Plants of Coastal First Peoples*. UBC Press and Royal British Columbia Museum.

Turner, Nancy J. 1997. *Food Plants of Interior First Peoples*. UBC Press and Royal British Columbia Museum.

Viereck, Eleanor G. 1987. *Alaska's Wilderness Medicines: Healthful Plants of the Far North*. Alaska Northwest Books, Portland, Oregon.

Walker, Marilyn. 1984. *Harvesting the Northern Wild*. Outcrop, Yellowknife, NT.

Walters, Dirk R. and David J. Keil. 1996. *Vascular Plant Taxonomy, 4th ed.* Kendall/Hunt Publishing Company, Dubuque, Iowa.

Ward-Harris, J. 1983. *More Than Meets the Eye: The Life and Lore of Western Wildflowers*. Oxford University Press, Toronto, Ontario.

Waterman, Catherine H. 1860. *Flora's Lexicon: The Language of Flowers*. Reprinted 2001, Algrove Publishing, Ottawa.

Wilkinson, Kathleen. 1999. *Wildflowers of Alberta: A Guide to Common and Other Herbaceous Plants*. University of Alberta Press and Lone Pine Publishing.

Wildflowers are perhaps
the most enchanting of all for me.
I love their delicacy, their disarming innocence
and their defiance of life itself.

Grace Kelly

Earth laughs in flowers.

Ralph Waldo Emerson

Index to Common and Scientific Names

M

N

O

P

Y

Z

I took a day to search for God
And found him not.
But where the scarlet lily flamed
I saw his footprint in the sod.

Bliss Carman

I envy not the rich their fields
Nor councillors their power,
While all the world my palace is
And every weed a flower.

Roundelay

Notes

...the genius of the earth, which is probably
that of the whole world, acts, in the vital struggle,
exactly as a man would act.
It employs the same methods, the same logic.
It attains its aim by the means which
we would use: it gropes, it hesitates,
it corrects itself time after time; it adds,
it suppresses, it recognizes and repairs its errors,
as we should do in its place.

Maurice Maeterlinck
The Intelligence of the Flowers, 1907

About the Author and Photographer

Alexandra Milburn is a writer with an affinity for wild plants. She comes to her study of wild plants not from an academic background, but from a personal interest and enthusiasm for the beauty, utility and tenacity of the wild things that many take for granted. Alexandra believes everyone's life can be enriched through identification and appreciation of the North's bounty of plants.

David Milburn is a physical geographer and hydrologist. Photography has been his hobby for over 30 years. David has taken many hundreds of photos of northern plants while accompanying Alexandra during her preparation of this book.

Georgie Milburn, their daughter, was also company along the trails. She has learned to recognize many local plants. The family set out several times a week to look for new species, and travelled far and wide in NWT in their quest.❦